BIO-DYNAMIC GARDENING AND FARMING

Of all occupations from which gain is secured, there is none better than agriculture, nothing more productive, nothing sweeter, nothing more worthy of a free man.
            Cicero, 104-43 B.C.

BIO-DYNAMIC GARDENING AND FARMING

ARTICLES
BY

EHRENFRIED PFEIFFER

MERCURY PRESS
SPRING VALLEY, NEW YORK

This edition of
Bio-Dynamic Farming and Gardening, Vol.I,
*first published by MERCURY PRESS in 1983,
is the first publication of Dr. Pfeiffer's
articles in this arrangement.
They were most recently published
in forty-four installments in
the periodical* Acres U.S.A.

*Permission for this publication
granted by Adelheid Pfeiffer.*

ISBN 0-936132-56-6
Copyright © 1983 by Mercury Press
All rights reserved.

Printed in the United States of America
MERCURY PRESS
Fellowship Community
241 Hungry Hollow Road
Spring Valley, N.Y. 10977

CONTENTS

	FOREWORD	page i
	ACKNOWLEDGMENTS	iii
CHAPTER I.	NEW DIRECTIONS IN AGRICULTURE	1
II.	DEFINITION OF STANDARDS	17
III.	THE ORGANIC CHEMICAL CONTROVERSY IN AGRICULTURE	27
IV.	THE DECLINE OF AMERICAN AGRICULTURE	47
V.	HOW TO BUY A FARM	53
VI.	SHALL WE PLOW?	61
VII.	THE BIO-DYNAMIC METHOD CAN PRODUCE QUANTITY TOO	65
VIII.	ORGANIC MATTER IN SOIL	73
IX.	HOW MOIST IS MOIST?	81
X.	MULCHING - RIGHT OR WRONG?	87
XI.	OBSERVATIONS OF BIO-DYNAMIC AND MINERAL TREATMENT OF SOIL	93
XII.	THE TREATMENT OF SOILS WITH REGARD TO HUMUS AND STRUCTURE	107
XIII.	SOIL PROFILES - A DIAGNOSTIC MEANS	119

FOREWORD

Despite the increasing number of publications on Bio-Dynamic gardening and farming, it seems appropriate to go back occasionally to the work of one of the great pioneers in this endeavor, Dr. Ehrenfried Pfeiffer. Therefore we present here the first of three volumes of articles by Dr. Pfeiffer.

A study of his publications shows not only the work of a pioneer in the attempts to rejuvenate our dying soils and to improve the nutrition of humanity, but actually the work of an expert whose efforts remain as valid today, 21 years after his passing, as they were then. One could say then that he was not only a pioneer, not only an expert, but also a prophet because much of what he foresaw and battled against is here today. Another attribute I would like to add is that of an inspired warrior battling to reclaim the earth for man. It is this last attribute which, to my mind, makes Dr. Pfeiffer's work seem so immediate; it imbues his style with a freshness, a vitality and an urgency which cannot but continue to inspire us to join in the necessary tasks.

I cannot resist to conclude here by quoting from the periodical *Acres U.S.A.*, which just finished publishing these same articles in 44 installments in order to expose "a new generation to the works of a giant with the realization that when we progress, we are all standing on the shoulders of giants." The series ends with the following editorial comment: "So ends the *Acres U.S.A.* presentation of The Pfeiffer Papers. These papers deserve a place in the library of every university in the country and on the bookshelf of every farmstead."

Gerald F. Karnow M.D.
Fellowship Community
Spring Valley, N.Y.

ACKNOWLEDGEMENTS

We would like to thank Adelheid Pfeiffer for granting Mercury Press permission to publish these papers in book form and we remember in gratitude the late Erica Sabarth who gathered them all with the hope that they would be published together.

We gratefully acknowledge the permission granted by the editors of the journal *Bio-Dynamics* to reprint the following chapters: Chapter III from Vol.X, No.3, 1952, chapter V from Vol.2, No.1, 1942, chapter VI from Vol.III, No.2, Winter 1944-45, chapter VII from No.59, 1961, chapter IX from Vol.XI, No.4, 1953, chapter X from No.51, 1959, chapter XI from Vol.X, N0.1, 1952, chapter XII from Vol.III, No.3-4, 1949, chapter XIII from Vol.XI, No.3, 1953.

I

NEW DIRECTIONS IN AGRICULTURE

In 1922/23 Ernst Stegeman and a group of other farmers went to ask Rudolf Steiner's advice about the increasing degeneration they had noticed in seed-strains and in many cultivated plants. What can be done to check this decline and to improve the quality of seed and nutrition? That was their question.

They brought to his attention such salient facts as the following: Crops of lucerne used commonly to be grown in the same field for as many as thirty years on end. The thirty years dwindled to nine, then to seven. Then the day came when it was considered quite an achievement to keep this crop growing in the same spot for even four or five years. Farmers used to be able to seed new crops year after year from their own rye, wheat, oats and barley. Now they were finding that they had to resort to new strains of seed every few years. New strains were being produced in bewildering profusion, only to disappear from the scene again in short order.

A second group went to Dr. Steiner in concern at the increase in animal diseases, with problems of sterility and the widespread foot-and-mouth disease high on the list. Among those in this group were the veterinarian Dr. Joseph Werr, the physician Dr. Eugen Kolisko and members of the staff of the newly established Weleda, the pharmaceutical manufacturing enterprise.

Count Carl von Keyserlingk brought problems from still another quarter. Then Dr. Wachsmuth and the present writer went to Dr. Steiner with questions dealing particularly with the etheric nature of plants, and with formative forces in general. In reply to a question about plant diseases, Dr. Steiner told the writer that plants themselves could never be diseased in a primary sense, "since they are the products of a healthy etheric world". They suffer rather from diseased conditions in their environment, especially in the soil; the causes of so-called plant diseases should be sought there. Ernst Stegeman was given special indications as to the point of view from which a farmer could approach his task, and was shown some first steps in the breeding of new plant types as a first impetus towards the subsequent establishment of the biological-dynamic movement.

In 1923 Rudolf Steiner described for the first time how to make the biodynamic compost preparations, simply giving the recipe without any sort of explanation - just "do this and then that". Dr. Wachsmuth and I then proceeded to make the first batch of preparation 500. This was then buried in the garden of the "Sonnenhof" in Arlesheim, Switzerland. The momentous day came in the early summer of 1924 when this first lot of 500 was dug up again in the presence of Dr. Steiner, Dr. Wegman, Dr. Wachsmuth, a few other co-workers and myself. It was a sunny afternoon. We began digging at the spot where memory, aided by a few landmarks, prompted us to search. We dug on and on. The reader will understand that a good deal more sweating was done over the waste of Dr. Steiner's time than over the strenuousness of the labour. Finally he became impatient and turned to leave for a five o'clock appointment at his studio. The spade grated on the first cowhorn in the very nick of time.

Dr. Steiner turned back, called for a pail of water, and proceeded to show us how to apportion the horn's contents to the water, the correct way of stirring it. As the author's walking stick was the only stirring implement at hand, it was pressed into service. Rudolf Steiner was particularly concerned with demonstrating the energetic stirring, the forming of a funnel or crater and the rapid changing of direction to make a whirlpool. Nothing was said about the possibility of stirring with the hand or with a birch-whisk. Brief directions followed as to how the preparation was to be sprayed when the stirring was finished. Dr. Steiner then indicated with a motion of his hand over the garden how large an area the available spray would cover. Such was the momentous occasion marking the birth-hour of a world-wide agricultural movement.

What impressed me at the time, and still gives one much to think about, was how these step-by-step developments illustrate Dr. Steiner's practical way of working. He never proceeded from preconceived abstract dogma, but always dealt with the concrete given facts of the situation. There was such germinal potency in his indications that a few sentences or a short paragraph often sufficed to create the foundation for a farmer's or scientist's whole lifework; the agricultural course is full of such instances. A study of his indications can therefore scarcely be thorough enough. One does not have to try to puzzle them out, but can simply follow them to the letter.

Dr. Steiner once said, with an understanding smile, in another very grave situation, that there were two types of people engaged in anthroposophical work: the older ones, who understood everything but did nothing with it, and the younger ones, who understood only partially or not at all, but immediately put suggestions into practice. We obviously trod the younger path in the agricultural movement, which did all its learning in the hard school of experience. Only now does the total picture of the new impulse given by Rudolf Steiner to agriculture stand clearly before us, even though we still have far to go to exhaust all its possibilities. Accomplishments to date are merely the first step. Every day brings new experience and opens new perspectives.

o   o   o

Shortly before 1924 Count Keyserlingk set to work in dead earnest to persuade Dr. Steiner to give an agricultural course. As Dr. Steiner was already overwhelmed with work, tours and lectures, he put off his decision from week to week. The undaunted Count then dispatched his nephew to Dornach, with orders to camp on Dr. Steiner's doorstep and refuse to leave without a definite commitment for the course. This was finally given.

The agricultural course was held from June 7 to 16, 1924, in the hospitable home of Count and Countess Keyserlingk at Koberwitz near Breslau. It was followed by further consultations and lectures in Breslau, among them the famous "address to Youth". I myself had to forego attendance at the course, as Dr. Steiner had asked me to stay at home to help take care of someone who was seriously ill. "I'll write and tell you what goes on at the course", Dr. Steiner said by way of solace. He never did get round to writing, no doubt because of the heavy demands on him; this was understood and regretfully accepted. On his return to Dornach, however, there was an opportunity to discuss the general situation. When I asked him whether the new methods should be started on an experimental basis, he replied: "The most important thing is to make the benefits of our agricultural preparations available to the largest possible areas over the entire earth, so that the earth may be healed and the nutritive quality of its produce improved in every respect; that should be our first objective. The experiments can come later". He obviously thought that the proposed methods should be applied at once.

This can be understood against the background of a conversation I had with Dr. Steiner en route from Stuttgart to Dornach shortly before the agricultural course was given. He had been speaking of the need for a deepening of esoteric life, and in this connection mentioned certain faults typically found in spiritual movements. I then asked, "How can it happen that the spiritual impulse, and especially the inner schooling, for which you are constantly providing stimulus and guidance bear so little fruit? Why do the people concerned give so little evidence of spiritual experience, in spite of all their efforts? Why, worst of all, is the will for action, for the carrying out of these spiritual impulses, so weak?" I was particularly anxious to get an answer to the question as to how one could build a bridge to active participation and the carrying out of spiritual intentions without being pulled off the right path by personal ambition, illusions, and petty jealousies; for these were the negative qualities Rudolf Steiner had named as the main inner hindrances. Then came the thought provoking answer: "This is a problem of nutrition. Nutrition as it is today does not supply the strength necessary for manifesting the spirit in physical life. A bridge can no longer be built from thinking to will and action. Food plants no longer contain the forces people need for this."

A nutritional problem which, if solved, could enable the spirit to become manifest and realize itself in human beings! With this as a background, one can understand why Dr. Steiner said that "the benefits of the bio-dynamic compost preparations should be made available as quickly as possible to the largest possible areas of the entire earth, for the earth's healing". This puts the Koberwitz agricultural course in proper perspective as an introduction to understanding spiritual, cosmic forces and making them effective again in the plant world.

In discussing ways and means of propagating the methods, Dr. Steiner said also that the good effects of the preparations and of the whole method itself were "for everybody, for all farmers" - in other words, not intended to the special privilege of a small, select group. This needs to be the more emphasised in view of the fact that admission to the course was limited to farmers, gardeners and scientists who had both practical experience and a spiritual-scientific, anthroposophical background. The latter is essential to understanding and evaluating what Rudolf Steiner set forth, but the bio-dynamic method can be applied by any farmer. It is important to point this out, for later on many people came to believe that only anthroposophists can practice the bio-dynamic method. On the other hand, it is certainly true that a grasp of bio-dynamic practices gradually opens up a wholly new perspective on the world, and that the practitioner acquires and applies a kind of judgment in dealing with biological - i.e. living - processes and facts which is different from that of a more materialistic chemical farmer; he follows nature's dynamic play of forces with a greater degree of interest and awareness. But it is also true that there is a considerable difference between mere application of the method and creative participation in the work. From the first, actual practice has been closely bound up with the work of the spiritual center of the movement, the Natural Science Section of the Goetheanum at Dornach. This was to be the source, the creative, fructifying spiritual element; while the practical workers brought back their results and their questions.

The name, "Bio-Dynamic Agricultural Method", did not originate with Dr. Steiner, but with the experimental circle concerned with the practical application of the new direction of thought.

In the Agricultural Course, which was attended by some sixty persons, Rudolf Steiner set forth the basic new way of thinking about the relationship of earth and soil to the formative forces of the etheric, astral and ego activity of nature. He pointed out particularly how the health of soil, plants and animals depends on bringing nature into connection again with the cosmic creative, shaping forces. The practical method he gave for treating soil, manure and compost, and especially for making the bio-dynamic compost preparations, was intended above all to serve the purpose of reanimating the natural forces which in nature and in modern agriculture were on the wane. "This must be achieved in actual practice," Rudolf Steiner told me. He showed how much it meant to him to have the School of Spiritual Science going hand-in-hand with real-life practicality when he spoke on another occasion of wanting to have teachers at the School alternate a few years of teaching (three years was the period mentioned) with a subsequent period of three years spent in work outside, so that by this alternation they would never get out of touch with the conditions and challenges of real life.

The circle of those who had been inspired by the agricultural course and were now working both practically and scientifically at this task kept on growing; one thinks at once of Guenther Wachsmuth, Count Keyserlingk, Ernst Stegemann, Erhard Bartsch, Franz Dreidax, Immanuel Voegele, M. K. Schwarz, Nikolaus Remer, Franz Rulni, Ernst Jakobi, Otto Eckstein, Hans Heinze, and of many others who came into the movement with the passing of time, including Dr. Werr, the first veterinarian. The bio-dynamic movement developed out of the co-operation of practical workers with the Natural Science Section of the Goetheanum. Before long it had spread to Austria, Switzerland, Italy, England, France, the north-European countries and the United States. Today no part of the world is without active collaborators in this enterprise.

The bio-dynamic school of thought and a chemically minded agricultural thinking confronted one another from opposite points of the compass at the time the agricultural course was held. The latter school is based essentially on the views of Justus von Liebig. It attributes the fact that plants take up substances from the soil solely to the so-called "nutrient-need" of the plant. The one-sided chemical fertilizer theory that thinks of plant needs in terms of nitrogen-phosphates-potassium-calcium, originated in this view, and the theory still dominates orthodox scientific agricultural thinking today. But it does Liebig an injustice. He himself expressed doubt as to whether the "N-P-K" theory should be applied to all soils. Deficiency symptoms were more apparent in soils poor in humus than in those amply supplied with it. The following quotation makes one suspect that Liebig was by no means the hardened materialist that his followers make him out to be. He wrote: "Inorganic forces breed only inorganic substances. Through a higher force at work in living bodies, of which inorganic forces are merely the servants, substances come into being which are endowed with vital qualities and totally different from the crystal." And further: "The cosmic conditions necessary for the existance of plants are the warmth and light of the sun." Rudolf Steiner gave the key to these "higher forces at work in living bodies and to these cosmic conditions." He solved Liebig's problem by refusing to stop short at the purely material aspect of plant-life. He went on, with characteristic spiritual courage and a complete lack of bias, to take the next step.

And now an interesting situation developed. Devotees of the purely materialistic school of thought, who once felt impelled to reject the progressive thinking advanced by Rudolf Steiner have been forced by facts brought to light during research into soil-biology to go at least one step further. Facts recognized as early as 1924 - 1934 in bio-dynamic circles - the significance of soil life, the earth as a living organism, the role played by humus, the necessity of maintaining humus under all circumstances, and of building it up where it is lacking - all this has become common knowledge. Recognition of biological, organic laws has now been added to the earlier realization of the undeniable dependence of plants upon soil nutrient substances. It is not too much to say that the biological aspect of the bio-dynamic method is now generally accepted; the goal has been perhaps overshot. But, important as are the biological factors governing plant inter-relationships, soil structure, biological pest control, and the progress made in understanding the importance of humus, the whole question of energy sources and formative forces -in other words, cosmic aspect of plant life - remains unanswered. The biological way of thinking has been adopted, but with a materialistic bias, whereas an understanding of the dynamic side, made possible by Rudolf Steiner's pioneering indications, is still largely absent.

Since 1924 numerous scientific publications that might be regarded as a first groping in this direction have appeared. We refer to studies of growth-regulating factors, the so-called growth-inducers, enzymes, hormones, vitamins, trace elements and bio-catalysts. But this groping remains in the material realm. Science has progressed to the point where material effects produced by dilutions as high as 1 : 1 million, no longer belong to the realm of the fantastic and incredible. They do not meet with the unbelieving smile that greeted rules for applying the biodynamic compost preparations, for these - with dilutions ranging from 1 to 10 to 1 to 100 million - are quite conceivable at the present stage of scientific thinking. Exploration of the process of photosynthesis - i.e. of the building of substance in the cells of living plants - has opened up problems of the influence of energy (of the sun, of light, of warmth and of the moon); in other words, problems of the transformation of cosmic sources of energy into chemical-material conditions and energies.

In this connection we quote from the book "Principles of Agriculture", written in 1952 by W. R. Williams, Member of the Academy of Sciences, USSR: "The task of agriculture is to transform kinetic solar energy, the energy of light, into the potential energy stored in human food. The light of the sun is the basic raw material of agricultural industry." And further: "Light and warmth are the essential conditions for plant life, and consequently also for agriculture. Light is the raw material from which agricultural products are made, and warmth is the force which drives the machinery - the green plant. The provision of both raw material and energy must be maintained. The dynamic energy of the sun's rays is transformed by green plants into potential energy in the material form of organic matter. Thus our first concrete task is the continuous creation of organic matter, storing up the potential energy of human life." And still further: "We can divide the four fundamental factors into two groups, according to their source: light and heat are cosmic factors, water and plant food terrestrial factors. The former group originates in interplanetary space . . ."

Or again: "The cosmic factors - light and heat - act directly on the plant, whereas the terrestrial factors act only through an intermediary (substance)."

We see that the author of this work rates knowledge of the interworking of cosmic and terrestrial factors as the first objective of agricultural science, while ranking organic substance (humus) second on the list of objectives of agricultural production. This is what was published in 1952. In 1924 Rudolf Steiner pointed out the necessity of consciously restoring cosmic forces to growth processes by both direct and indirect means, thereby freeing the present conception of plant nature from a material, purely terrestrial isolation; only through such restoration would it be possible to re-energize those helpful and constructive forces capable of halting degeneration. He said to me: "Spiritual scientific knowledge must have found its way into practical life by the middle of the century if untold damage to the health of man and nature is to be avoided."

o o o

Our research work began with the attempt to find reagents to the etheric forces and to discover ways of demonstrating their existence. Suggestions were given which could only later be brought to realization in the writer's crystallization method. Then it was our intention to proceed to expose the weak points in the materialistic conception and to refute its findings by means of its own experimental methods. This meant applying exact analytical methods in experimentation with physical substances, and even developing them to a finer point. We proposed to work quantitatively as well as qualitatively. During my own years at the University, for example it was my regular practice to lay my proposed course of studies for the new term before Rudolf Steiner for guidance in the choice of subjects. On one occasion he urged me to take simultaneously two - no, three - main subjects, chemistry, physics and botany, each requiring six hours a day. To the objection that there were not hours enough in the day for this, he replied simply, "Oh, you'll manage it somehow."

Again and again, he steered things in the direction of practical activity and laboratory work, away from the merely theoretical.

Suggestions of this kind were constantly in my mind during the decades of work which arose from them. They led me not only to work in laboratories, but also to apply the fundamentals of this new outlook to the management of agricultural projects, both in a bio-dynamic and in an economic sense. Dr. Steiner had insisted on my taking courses and attending lectures in political economy as well as in science, saying, "one must work in a businesslike, profit-making way, or it won't come off". Economics, commercial history, industrial science, even mass-psychology and other such subjects were proposed for study, and when the courses were completed, Dr. Steiner always wanted a report on them. On these occasions he not only showed astounding proficiency in the various special fields, but - what was more surprising - he seemed quite familiar with the methods and characteristics of the various professors. He would say, for example, "Professor X is an extremely brilliant man, with wide-ranging ideas, but he is weak in detailed knowledge. Professor Z is a silver-tongued orator of real elegance. You needn't believe everything he says, but you must get a thorough grasp of his method of presentation."

From these and many other suggestions it was clear what had to be done to promote the bio-dynamic method. There was the big group of practicing farmers, whose task it was to carry out the method in their farming enterprises, to discover the most favorable use of the preparations, to determine what crop rotations build up rather than deplete humus, to develop the best methods of plant and animal breeding. It took years to translate the basic ideas into actual practice. All this had to be tried out in the hard school of experience, until the complete picture of a teachable and learnable method, which any farmer could profitably use, was finally evolved. Problems of soil treatment, crop rotation, manure and compost handling, time-considerations in the proper care and breeding of cattle, fruit tree management and many other matters could be worked out only in practice through the years.

Then there was the problem of coming to grips with agricultural science. Laboratories and field experiments had to provide facts and observational material. I was now able to profit from the technical and quantitative-chemical education urged upon me by Dr. Steiner. This was the sphere in which the shortcomings and weaknesses of the chemical soil-and-nutrient theory showed up most clearly, and where today - after more than thirty years - one can see possibilities of building a bridge between recognition of the existence of cosmic forces and exact science.

The first possibility of breaking through the hardened layer of current orthodox opinion came through discoveries that cluster around the concept of the so-called trace elements. Dr. Steiner had pointed out as early as 1924 the existence of these finely dispersed material elements in the atmosphere and elsewhere, and had stressed the importance of their contribution to healthy plant development. But it still remained an open question whether they were absorbed from the soil by roots or from the atmosphere by leaves and other plant organs. In the early thirties, spectrum analysis showed that almost all the trace elements are present in the atmosphere in a proportion of 10 to the -6 to 10 to the -9. The fact that trace elements can be absorbed from the air was established in experiments with *Tillandsia usneodis*. It is now common practice in California and Florida to supply zinc and other trace elements, not via the roots, but by spraying the foliage, since leaves absorb these trace elements even more efficiently.

It was found that one-sided mineral fertilizing lowers the trace-element content of soil and plants, and - most significantly - that to supply trace elements by no means assured their absorption by plants. The presence (or absence) of zinc in a dilution of 1 to 100 million decides absolutely whether an orange tree will bear healthy fruit. But in the period from 1924 to 1930 the biodynamic preparations were ridiculed "because plants can not possibly be influenced by high dilutions".

Zinc is singled out for mention here not only because treatment with very high dilutions of this trace element is especially essential for both the health and the yield of many plants, but also because it is an element particularly abundant in mushrooms. A comment by Rudolf Steiner indicates an interesting connection which can be fully understood only in the light of the most recent research. We read on page 107 of the Agricultural Course: "...Harmful parasites always consort with growth of the mushroom type..... causing certain plant diseases and doing other still worse forms of damage..... One should see to it that meadows are provided with mushrooms ... Then one can have the interesting experience of finding that where there is even a small growth of mushrooms in a meadow near a farm, the mushrooms (fungi) owing to their kinship with bacteria and parasites, keep them away from the farm. It is often possible by implanting mushrooms in meadows to keep in this way off all other pests."

Organism of the fungus type include the so-called fungi imperfecti and a botanical transition form, the family of actinomycetes and streptomycetes, from which certain antibiotic drugs are derived. I have found that these organisms play a very special role in humus formation and decay, and that they are abundantly present in the bio-dynamic manure and compost preparations. The preparations also contain an abundance of many of the most important trace elements, such as molybdenum, cobalt, zinc and others whose importance has been experimentally demonstrated.

Now a peculiar situation was found to exist in regard to soils. Analyses of available plant nutrients showed that the same soil tested quite differently at different seasons. Indeed, tests showed not only seasonal but even daily variations. The same soil sample often disclosed periodic variations greater than those found in tests of soils from adjoining fields, one of which was good, the other poor. Seasonal and daily variations are influenced, however, by the earth's relative position in the planetary system; they are, in other words, of cosmic origin. It has actually been found that the time of day or the season of the year influences the solubility and availability of nutrient substances. Numerous phenomena to be observed in the physiology of plants and animal (e.g., glandular secretions, hormones) are subject to such influences. The concentration of oxalic acid in bryophyllum leaves rises and falls with the time of day with almost clock-like regularity. Although in this and many other test cases the nutrients on which the plants were fed were identical, the increase or decrease in the plant's substantial content varied very markedly in response to varying light-rhythms and cycles. Joachim Schultz, a research worker at the Goetheanum whose life was most unfortunately cut short, had begun to test Dr. Steiner's important indication that light activity acts with growth-stimulating effect in the morning and late afternoon hours, while at noon and midnight its influence is growth-inhibiting.

When I inspected Schultz's experiments, I was struck by the fact that plants grown on the same nutrient solution had a wholly different composition according to the light-rhythms operative. This was true of nitrogen, for example. Plants exposed to light during the morning and evening hours grew strongly under the favorable influence of nitrogen activity, whereas if exposed during the noon hours, they declined and showed deficiency symptoms. The way was thus opened for experimental demonstration of the fact that the so-called "cosmic" activity of light, of warmth, of sun forces especially, but of other light-sources also, prevails over the material processes. These cosmic forces regulate the course of material change. When and in what direction this takes place, and the extent to which the total growth and the form of the plant are influenced, all depend on the cosmic constellation and the origin of the forces concerned. Recent research in the field of photosynthesis has produced findings which can hardly fail to open the eyes even of materialistic observers to such processes. Here, too, Rudolf Steiner is shown to have been a pioneer who paved the way for a new direction of research. It is impossible in an article of this length to report on all the phenomena that have already been noted, for they would more than fill a book. But it is no longer possible to dismiss the influence of cosmic forces as "mere superstition" when the physiological and biochemical inter-relationships of metabolic functions in soil life, the rise and fall of the sap in the plant, and especially processes in the root-sphere are taken in consideration.

o   o   o

In an earlier view of nature, based partly on old mystery-tradition and partly on instinctive clairvoyance - a view originating in the times of Aristotle and his pupil Theophrastus, and continuing on to the days of Albertus Magnus and the late medieval "doctrine of signatures" - it was recognized that relationships exist between certain cosmic constellations and the various plant species. These constellations are creative moments under whose influence species became differentiated and the various plant forms came into being. When one realizes that cosmic rhythms have such a significant influence on the physiology of metabolism, of glandular functions, of the rise and fall of sap and of sap pressure (turgor), only a small step remains to be taken by conscious future research to the next realization which will achieve an experimental grasp of these creative constellations. Many of Rudolf Steiner's collaborators have already demonstrated the decisive effects of formative forces in such experiments as the capillary tests on filter paper of L. Kolisko and the plant and crystallization tests of Pfeiffer, Krueger, Bessenich, Selawry and others.

Rudolf Steiner's suggestions for plant breeding presented a special task. Research in this field was carried out by the author and other fellow-workers (Immanuel Voegele, Erika Riese, Martha Kuenzel and Martin Schmidt), either in collaboration or in independent work. Proceeding from the basic concept of creative cosmic constellations, one can assume that the original creative impetus in every species of sub-type slowly exhausts itself and ebbs away. The formative force of this original impulse is passed on from plant to plant in hereditary descent by means of certain organs, such as chromosomes. One-sided quantity-manuring gradually inhibits the activity of the primary forces, and results in a weakening of the plant. Seed quality degenerates. This was the initial problem laid before Rudolf Steiner, and the bio-dynamic movement came into being as an answer to it.

The task was to reunite the plant, viewed as a system of forces under the influence of cosmic activities, with nature as a whole. Rudolf Steiner pointed out that many plants which had been "violated", in the sense of having been estranged from their cosmic origin, were already as far gone in degeneration that by the end of the century their propagation would be unreliable. Wheat and potatoes were among the plant types mentioned, but other grains such as oats, barley and lucerne belong to the same picture. Ways were sketched whereby new strains with strong seed forces could be bred from "unexhausted" relatives of the cultivated plants. This work has begun to have success; the species of wheat have already been developed. Martin Schmidt carried on significant researches, not yet published, to determine the rhythms of seed placement in the ear, and to show in particular the difference between food plants and plants grown for seed. According to Rudolf Steiner, there is a basic difference between the two types, one of which is sown in autumn, nearer to winter, and the other nearer to summer. Biochemists will eventually be able to confirm these differences materially in the structure of protein substances, amino-acids, phospholipids, enzyme-systems and so on by means of modern chromatographic methods.

The degeneration of wheat is already an established fact. Even where the soil is good, the protein content has declined; in the case of soft red wheat, protein content has sunk from 13% to 8% in some parts of the United States. Potato growers know how hard it is to produce healthy potatoes free from virus and insects, not to mention the matter of flavor. Bio-dynamically grown wheat maintains its high protein level. Promising work in potato breeding was unfortunately interrupted by the last war and other disturbances.

Pests are one of the most interesting and instructive problems looked at from the bio-dynamic viewpoint. When the biological balance is upset, degeneration follows; pests and diseases make their appearance. Nature herself liquidates weaklings. Pests are therefore to be regarded as nature's warning that the primary forces have been dissipated and the balance sinned against. According to official estimates, American agriculture pays a yearly bill of five thousand million dollars in crop losses for disregarding this warning and another seven hundred and fifty million dollars on keeping down insect pests. People are beginning to realize that insect poisons fall short of solving the problem, especially since the destruction of some of the insects succeeds only in producing new, more resistant kinds. It has been established by the most advanced research (Albrecht of Missouri) that one-sided fertilizing disturbs the protein-carbohydrate balance in plant cells, to the detriment of proteins and the layer of wax that coats plant leaves, making the plant "tastier" to the insect predators. It has been a bitter realization that insect poisons merely "preserve" a part of moribund nature, but do not halt the general trend towards death. Experienced entomologists, who have witnessed the failure of chemical pest control and the threat to health associated with it, are beginning to speak out and demand biological controls. But according to the findings of one of the American experimental stations, biological controls are feasible only when no poisons are used and an attempt is made to restore natural balance. In indications given in the agricultural course, Rudolf Steiner showed that health and resistance are function of biological balance, coupled with cosmic factors. This is further evidence of how far in advance of its time was this spiritual-scientific, Goethean way of thought.

The author is thoroughly conscious of the fact that this exposition touches upon only a small part of the whole range of questions opened up by Rudolf Steiner's new agricultural method. He is also aware that other collaborators would have written quite differently, and about different aspects of the work. These pages should therefore be read in accordance with their intentions: as the view from a single window in a house containing many rooms.

II
DEFINITION OF STANDARDS

1. WHAT IS ORGANIC?

The term implies that the life process is the origin of the organic product; that the product derives from natural growth of a plant, microorganism or animal. It also implies that this process has been carried out in an orderly fashion according to the laws of nature. Health can only stem from health. If a farmer produces a food plant or animal, he produces the maximum health value if all factors related to the growth process are healthy; that is, soil, plant, animal. No health food can derive from a deficient soil, a deficient plant or a diseased plant or animal.

The term "organic" in the strictly chemical definition means a compound of the carbon chemistry as differentiated from a mineral, inorganic substance. The organic substance based on the chemical definition, can be natural as it exists in plants, animals, or the human body. It can also be a synthetically produced substance as the chemist produces it in the laboratory or industry. To the chemist, it is "organic" as long as it is a carbon compound. Carbon compounds can be many; it is estimated that today some six million different compounds have been described and are known, while all other chemical elements together have formed only some 50,000 to 400,000 compounds.

Our definition of "organic" is not the one of the chemist, but pertains to the modus of production by nature, not artificially or synthetically. It is felt that Nature can transmit a quality (of life) which the synthetic product misses. In certain cases this difference becomes quite apparent. For instance, in nutrition, vitally important substances such as vitamins, amino acids, hormones, sugars show optical rotation in polarized light (left or right rotary), while synthetic substances are optically inactive (racemic). Many chemists deny that a substance can be different, whether natural or synthetic, as long as it has the same chemical formula. Biologically and physiologically, however, there may be differences in behavior. Testing methods, bio-assays, need to be developed to demonstrate the difference.

2. WHAT IS NATURAL FOOD?

It is quite obvious that Nature operates according to strictly observable laws. It organizes, directs functions to their proper place at the right time. In the plant and animal organism, a system of organs and their functions brings about the desired result in a concerted effort. Health exists if all functions are properly coordinated, controlled, by an organizing factor in a harmonious way. Nature always tends to produce a balance. The organic concept has been expressed already almost two thousand years ago by the words of St. Paul (I.Cor.12):"For the body is not one member but many ..... but now are they many members, yet one body."

It is the supreme task of the scientist who follows the organic concept to demonstrate the unity or totality (wholeness) of functions and of the growers to apply the principles to the benefit of Nature's and human health. To work with Nature and not against it is the aim of the N.F.A. affiliated grower. To lose oneself in any one detail would only mean incompleteness and therefore a cause of disease and breakdown.

3. WHAT ARE NATURAL FERTILIZERS?

The answer to this question has led to many discussions. It is necessary to keep concepts quite clear with regard to this matter. The organic farming and gardening movement has been steadily growing over the last 15 years. We have to differentiate between organic and mineral fertilizers. Lime, rock phosphate, green sand and potash rock are minerals. Nevertheless, they can be used according to the concept of the organic movement for they are natural. To call rock phosphate or potash rock "organic" is a misnomer. Superphosphate is a processed mineral fertilizer. Organic fertilizers are: manure of all kinds, compost, peat, guano. Manures are "processed" by the digestive action of the animal which produces it. Compost is "processed" by fermentation, by way of micro-organisms and the enzymes which they produce by biological breakdown. They are, nevertheless, "natural", for it is nature's bio-chemistry which breaks the crude or raw organic materials down to compost. The aim of all composting processes is to produce, finally, "humus" that state of organic matter in soil which is stabilized and maintains soil life. There are many degrees of breakdown between the raw offal and the biologically digested end product. The nearer to the state of humus, the more valuable the compost is for the production of soil health. The cruder, the more the danger that undesirable by-products and by-effects show up. Scatol and indol are "organic", but are quite poisonous compounds in fresh manure or feces. If they enter by way of insufficiently decomposed materials into the soil, they can be absorbed by plants and transmit their off-flavors to the vegetables - for instance, cabbage, sprouts, potatoes, etc; that is, produce inferior quality.

In mineral fertilizers the availability of the contents decides their value; whether nitrogen, phosphate, potash, lime (the big four) or the minor elements, magnesium, manganese or strictly trace minerals. Neither the formula (NPK) on the bag nor the statement on the label that the contents are available means that the minerals will remain available once they hit the soil. In imbalanced soils without soil life, that is, under adverse conditions, the mineral components may become unavailable and therefore insufficient. Organic matter and soil life are a must to maintain availability.

Organic fertilizers are valuable to the extent that they are "digested" by microlife, teeming with microlife and enhancing the growth of microlife in the soil. This, in turn, will increase the availability of the mineral elements. They support, therefore, the natural process. Urea is chemically speaking, organic, but it is a product of industry and therefore not natural. Only the urea which is contained in manure is natural.

Finally, a word about sewage sludge. This has to be completely broken down to be most beneficial. Nothing of the original nature of feces should be left; otherwise, the danger of pathological organisms as well as of toxic by-products, exists.

4. ARE PESTICIDES AND INSECTICIDES NATURAL AIDS IN THE PRODUCTION OF CROPS?

It is, alas, true that insect pests and fungus diseases are abundant; in fact, have tremendously increased with the "progress" of agriculture. Their presence is a sign of the disturbed balance of Nature. Their increase indicates that fundamental laws of growth, soil, and plant health have been violated. Just to "kill the insect" does not correct the violation of Nature's balance which has brought about the increase of pests. Official estimates state that the damage done to U.S.A. crops per year amounts to a value of a billion dollars.

Insecticides and pesticides fall into two groups:
(a) Chemical poisons such as DDT, chlordane, lyndane, chlorinated hydrocarbons, fluorinated and other so-called economic poisons. Their number goes into the thousands, and every year numerous new compounds are added. This means that the ideal solution has not been found as yet. As long as the fundamental laws of Nature's balance have been violated, the goal can never be reached and all measures are only half measures. Some of the insecticides are looked upon with suspicion because insects have become immune and stronger poison is needed to interrupt the vicious cycle. The case of DDT is best known. Some of these insecticides remain on fruit, vegetables, and can accumulate to constant toxic levels (excepting the immunized insects). They form residues which eventually will reach the human consumer. Protective laws are already necessary stating the tolerance levels of such insecticides. The consumer who is aware of the benefits of uncontaminated, natural food, wants to know whether an agricultural product has been sprayed or not. We consider it only fair practice that the labeling laws are extended to a declaration of the sprays which were used. The consumer then can make his choice.
(b) Organic natural poisons which leave no residue and do not build up immunities. Such poisons deteriorate in the soil or on the leaves and fruit within a short time and are, therefore, harmless in the long run. Most of these poisons are extracted from plants and therefore merit the term "natural". Hereto belong Derris, Rotenone, Pyrethrum, Quassia,

Nicotine, Rhyania. This writer would tolerate these organic, natural poisons inasmuch as their residues, if any, would reach the consumer. If the last application is done four weeks prior to harvest, there is reasonable chance that there are no residues left.

Some growers are in the fortunate situation that they are not bothered with pests. These are very few, though. The majority still have to combat pests. Especially in the conversion phase to the organic and natural production method, they will have problems. If they happen to live in an otherwise highly contaminated area, the problems are still worse. From these growers the benefits of natural, organic poisons should not be withheld by a statement: "We don't want any sprays whatsoever." They should be given a chance to convert and gradually establish natural balance and biological control.

The consumer again is entitled to know how his food has been grown, and we feel it is only fair to him to tell on the label "unsprayed" or "sprayed with such-and-such organic, natural poison". He then can make his own choice.

5. WHAT IS NATURAL BALANCE?

It has already been mentioned that the maintenance of balance is necessary with regard to fertilizers, soil; that there is a biological balance which effects the control of insect pests. In fact, the problem of balance holds the key position to all life and health. The human body maintains a salt and water balance, a blood sugar balance, hormone balance. There are many intricate processes, enzymatic reactions which are directed in such a way that the balance is maintained. Likewise, in the plant realm and in soils all factors of life are integrated. Modern cultivation methods geared toward high quantity production have frequently disturbed the equilibrium and reduced quality production and with it have reduced the food and health value. In the striving for high average yields of wheat, for instance, the balance has been shifted; the protein content and baking quality of the wheat flour was reduced in favor of a higher carbohydrate content. This, in turn, has made the plant more susceptible to attacks by pests. For decades the farmer was told NPK fertilization is needed and the important trace minerals were neglected. This way, deficiencies have been caused. That deficiency symptoms show up always means that Nature has been one-sidedly treated. Numerous vitamin preparations and supplements are needed nowadays in order to correct these deficiencies. The demand of the day is to restore natural health and nutritious value first and then to see whether any correction is still needed.

Every engineer knows that each construction material has a load limit beyond which it will break. So the entire growth process has its limitation. It is a necessity that science begins to realize the need for a deeper insight into the checks and balances and learns to work with instead of against Nature. Farming practices need to be amended accordingly.

6. IS COMMERCIAL FERTILIZER OBJECTIONABLE?

It is evident that any fertilizer, whether mineral or organic, is commercial as soon as it is produced and sold. The manure or compost which the farmer or gardener produces for his own use is not commercial, but nevertheless has an economic value.

Commercial mineral fertilizers, according to law, have to be declared according to their formula and contents. It is only fair if commercial organic fertilizers are likewise labeled with regard to formula and contents. However, not only NPK determines the value of a fertilizer. Trace minerals play an important role. The effects of organic matter, humus, and the intrinsic value of it should also find proper and adequate evaluation. This is frequently not done. After all, it is not the formula on the bag that decides the issue, but how much of the fertilizer elements remain available for soil life and plant production. Organic soils with the proper nutrient balance maintain themselves much better and are less subject to erosion, washing out and leaching than run-down, one-sided mineralized soils.

7. WHAT IS A FERTILE SOIL?

This writer once analyzed two different soils several years ago which showed almost the same contents. One soil produced a top yield, the other produced a poor crop. Apparently there are other factors besides the chemistry which determine whether a soil is fertile or not. One important factor is the water absorption and water holding capacity of a soil. It is much too little realized that water is the limiting factor with regard to crop yield. Some desert soils in the analysis show sufficient amounts of phosphate and potash; they are not fertile because they lack water and organic matter. Some soils, richly fertilized, may contain all nutrients, plants set out to a good start and suddenly stop growing. A hardpan underneath prohibits the roots from growing deeper and contains a toxic accumulation of salts and acidity. The bacterial life in the soil decides the availability of minerals, the rate of nitrification and of useful nitrogen reserve. The most favorable bacteria life can only exist if the drainage is well maintained.

A root consumes its own weight by way of oxygen per day. Waterlogged soil increases acidity and unfavorable reactions. Spectacular results can be seen by just improving the soil structure (so-called physical conditions versus chemical nutrients). A fertile soil is one which is well supplied with available minerals, humus or organic matter better than 2 %, water holding capacity, good drainage, good aeration, favorable structure conditions, good tilth, proper cultivation, and balance of rotation versus exhausting monoculture. It is a grave error, frequently encountered, to think that fertilizer alone can amend all problems of the soil. We are frequently asked: "What can we throw on to get better crops?" The question should be: "How can we introduce such measures as to make the available nutrients more effective?" A prayer hewn in stone at the cathedral in Chester, England, says: "O Lord, give me a good digestion but also something to digest." The same can be said with regard to soils: something to digest (soil nutrients, water, air, organic matter, soil life) - but also a good digestion; that is, a life process properly regulated so as to make the best use of what the soil contains. The general term is: to use soil building and conserving practices. Fertilizer is only a small item of a manifold pattern. Only the integration of all factors will secure health and nutritional value of crops.

Under deficient and unbalanced conditions, pests will find an easy prey. The spraying of toxic minerals alone does not solve the problem. Soil building practices have to support the combat against pests. A farmer or grower who starts out on a run-down condition cannot expect to see results right away. There is a transitional period during which the soil is rebuilt. We do not want the grower to go out of business until the soil has recovered. But we suggest to use natural means, for instance, sprays from not-lasting plant poisons, produced by Nature to bridge over a period of years until resistances have been built up. The breeding of smut or blight resistant varieties is an important part of this program. If a larger area is badly infested and neighboring farms are intensively treated with economic poisons, the chances are that the insects will move into the unsprayed area at first.

The biological control of citrus orchards has been very efficient but requires that no detrimental sprays are used which handicap the introduction of predatory insects. Usually the grower who introduces biological control takes a beating during the first year of conversion. Lasting toxic effects have to weather away too. All this makes it necessary that a transitional period is permitted until the soil is rebuilt and the resistance of plants increased. This period may be from one to four years. While the aim is to grow healthy fruit and vegetables without sprays, one should not become dogmatic about it, but give due consideration to the circumstance.

A word should be said about the customer too. As long as he buys only size, color and polish (or even only beautiful wrappings and print), there is no incentive for the grower to improve quality. The customer should become quality conscious and demand unsprayed fruit and pay a premium price for it. Soon progressive growers will become alert and learn to improve their production methods.

Quality cannot always be discovered by the outer look. An apple might look fine on the outside and still have brown rot (boron deficiency) in the core. The major and most important nutrients (amino acids, for instance) are located in the skin of the apple and potato, in the pulp and skin of the orange. For health reasons, one would like to eat the skins were it only free of spray residues.

Nature produces a protective layer of waxes on many leaves, fruits when the soil system is balanced. This protective layer is frequently missing under one-sided fertilizer conditions. No washing and afterwards applying of an oil or paraffin spray for protection (for storage and shipping) can substitute for the beneficial effect of the natural waxes. Why plant breeders have never paid attention to producing plants with said protective layer is beyond our comprehension.

Finally, a word about protection for shipping and storage. Grapes, dates, raisins, plums are said not to keep well unless protected by sulfurdioxide and bromethylene spray or gas. It is true that most of this can be removed again by aeration and washing, but particles of the protective agent will be absorbed. Fruit and vegetables grown under balanced conditions have proven to be of better carrying quality.

III

THE ORGANIC-CHEMICAL CONTROVERSY IN AGRICULTURE

Many articles have appeared recently discussing the use of chemical versus the organic method in agriculture. The general tendency in all these publications is to express appreciation of certain of the values of organic matter but to criticize the organic movement as a whole as fanatical, mystical, and to state that this country has about doubled its acre yields with chemical fertilizers while there would not be enough organic material (manure, compost, green manuring) available to supply the needs of a high production agriculture. One particular criticism of the organic movement that is often voiced is that the 'fanatics have claimed that chemical fertilizer poisons the soil and spray residues on fruits and vegetables are a menace to public health.' It is also frequently stated that scientific tests have shown no difference in yield and in vitamin production in plants, whether grown organically or chemically. It has even been shown that plants can grow, produce fruit, and contain all necessary mineral elements, proteins and vitamins if grown in properly adjusted nutrient solutions without any soil at all.(1,2)

Occasional emotional outbursts on both sides of the argument can be left out of the picture. They help neither cause. We need only take an objective stand in making a mental inventory of present day knowledge.(3)

This writer has spent more than 25 years in the study of organic matter in soils, has done laboratory as well as field research, has practiced farming and consulted with other farmers and growers. He has proved, in making a living on his own farms, that his point of view works. He speaks out of practical experience and does not indulge in theories. He does not take part in some of the outbursts of the organic movement for he is a scientist for whom only facts count.(4) Nor does he share in the attitude of the other side that chemical fertilizers will solve all the problems of survival. Somewhere between the two there must exist a workable path which must be sought for the sake of the maintenance of fertile soils. The writer has tried, therefore, to apply scientific methods to the organic idea. This is what he has found:

The defenders of the organic method are enthusiasts not out of sheer intellectual delight but because they have tried 'something' in the garden and field and have observed results which they could not obtain otherwise. They have had troubles with their soils which have disappeared. One cannot consequently blame them for their enthusiasm. They would, however, do well to report only facts and findings and omit enthusiastic and critical statements or opinions. The facts speak for themselves. Then it should also be realized that many statements made years ago in favor of organic matter have become common knowledge today, even among the orthodox, and it should be acknowledged that the value and importance of organic matter is no longer doubted. To fight against science as it was years ago is really ridiculous.(5)

It should be acknowledged by both camps that not all the problems have been solved. Open and unprejudiced discussion of the unsolved problems will lead to the fruitful path.

There is, for instance, the organic school's statement that insect pests can be overcome when no inorganic, chemical fertilizers are used. This writer has found insects and fungi diseases in organically treated orchards, gardens and fields, but he has also observed that they were reduced to a bearable minimum and did not spread like wildfire. On the other side he has also seen some fertilizer treated land relatively free of pests. However, to pretend that sprays and economic poison have licked the problem is a far cry from reality.

Dean Stanley B. Freeborn of the U.C. College of Agriculture reported recently to the National Agricultural Chemical Association in San Francisco: 'Despite the use of chemicals, the annual loss caused by insect pests is about $4 billion; losses due to fungi and other plant diseases is another $4 billion and weed damage about $5 billion, a total of $13 billion. Without the use of chemicals it would have been vastly more.'

No wonder the farmer gets worried, searches, and begins to doubt that his present methods are the sure cure. It costs him a lot of money to combat diseases and pests and frequently he spends all that he gains in dollars in yield in order to maintain this yield. Insecticides and weed killers have done wonders. But they have also backfired. Insects have become immune to DDT and other insecticides, the natural enemies of destructive insects have been killed and the pests - after overcoming the first impact of the new insecticide - have developed more vigorously than ever. The trend now is toward biological control in certain areas without insecticides, but this necessitates the restoration of natural balances as the first step. It is true that the skillful use of fertilizers has increased yields.(6) On the other hand, it has been observed that soil structures have declined; hardpan, crusted and impenetrable surface soils have appeared; in general, unfavorable soil structures have developed. Fertilizers have washed out and become unavailable. There is, for instance, the statement by H. D. Chapman of the Riverside Citrus Experiment Station, *Progress Made in Tree Feeding* (Citrus Leaves, Feb. 1952), reporting that in lysimeter tests over a period of 15 years, 3,552 pounds of potassium, 448 pounds of nitrogen, 227 pounds of phosphorus were lost in spite of a cover crop of mustard during the winter and in spite of the fact that each year calcium nitrate at the rate of 200 lbs/acre was added. The pH changed from 6.6 to 7.9 during this period, and chlorine increased to 119 pounds and sodium to 1913 pounds, both undesirable.

The onesided use of fertilizers geared to provide only NPK has disturbed the balance of soils and created trace mineral deficiencies. Through exclusive emphasis on NPK over many years, research as well as agricultural practice has neglected the importance of biocatalysts in plant growth. Scientists are now in a hurry to catch up.

Phosphate fertilizers have become unavailable. Radioactive tracer studies have shown that probably only 2 to 10% of the phosphate applied will remain available. The rest is locked up in the soil, may be washed down into deeper layers inaccessible to plant roots. Whether these fertilizers will act as a reserve and can be brought back into available forms nearer the surface depends entirely upon the proper treatment of the soil. It will certainly not become available if crusted soils and hardpan prevail. (The increase of organic matter counteracts the process of increasing unavailability.) It is now acknowledged that deep-rooting weeds play an important and beneficial role in this regard. Alkaline soils lock phosphate up completely. While in acid soils the use of natural rock phosphates as recommended by the organic school can be beneficial, in alkaline soils the rock phosphate may become and remain unavailable unless there is such microlife in the soil as to render it available.

Similar conditions exist as regards potash.(7) There may be tremendous amounts of potash bound up in the soil but these are not available and it is the farmer's task again to increase and mobilize microbial action in order to make these potash reserves accessible. On the other hand, water soluble potash fertilizers may be washed out in areas of high rainfall or intensive irrigation and be completely lost. Fertilizer legislation has helped to increase losses of valuable 'plant food' by emphasizing the use of available phosphate and potash in fertilizers. This emphasis was based on the old concept of availability and water solubility apparently making these fertilizers readily accessible. However, the fact that these fertilizers become unavailable or are washed out at once when they encounter certain soil conditions was not taken into consideration. Organic matter again, by means of its high absorption factor and retentive capacity would play a beneficial part by holding the fertilizer in exactly that surface layer where it can do its best for the growing roots. Organic matter, with its microlife, allows a slow but steady release of important minerals and cuts down excessive losses and washing out.

Applying fertilizers to soils low in organic matter or where they can be locked up or washed out can become a rather wasteful practice. Many farmers have observed that they have to use more and more from year to year in order to maintain yields. At the same time they are fighting hard against increasingly unfavorable soil structures. They really want to know how to make less fertilizer go farther. This is what the much criticized organic farmers have now observed that they can do. Hence their enthusiasm. Organic matter feeds the microlife of soils. Soil bacteria in turn make minerals available and increase organic matter in soil. Organic matter is the savings account of the soil. It stores moisture and maintains this moisture longer during periods of drought. On the other hand, it readily absorbs water and reduces the danger of 'wet feet' for plant roots which may cause root rot and fungus infection.(8)

One acre of surface soil weighs approximately two million pounds. An organic matter content of 2% amounts to 40,000 pounds of organic matter of all varieties. On the average, 5% of the organic matter is nitrogen, or 2,000 pounds of organic nitrogen can be present at the 2% organic matter level. Microorganisms are the richest source of nitrogen since their bodies contain as much as 10 to 12% nitrogen. At a 1% organic matter level the organic nitrogen reserve is 1,000 pounds, at a 3% level the reserve is 3,000 pounds, and so on. This reserve can easily be lost. It can also be maintained. Humus which is the 'digested' form of organic matter and rather a 'state' of matter in soil than a chemical, is stable when proper methods of cultivation are used and the microorganisms and their life cycle are maintained. To have 2% organic matter does not necessarily mean that there is 2% humus present. But a high microlife in soil with a soil reaction between pH 6.0 and 7.5 and proper aeration of soil indicates a high humus level. In fact, the more aerobic soil microorganisms are present the higher will the humus fraction be. The carbon-nitrogen relationship of 10:1 or 11:1 is the ideal one and exists in humus.

'Humus is usually in a high dynamic condition, for it is constantly being formed from plant and animal residues and is continuously being decomposed by microorganisms. It serves as a source of energy for various groups of microorganisms...Humus is highly colloidal...Humus is characterized by a high cation-exchange capacity; it combines with various inorganic soil constituents; it absorbs large quantities of water...it has already been shown that soil humus is an important factor in the control of aeration, water holding capacity and granulation of field soils. Humus possesses other physico-chemical properties which make it a highly valuable soil constituent.'

These statements (taken from *Fundamentals Of Soil Science* by C. E. Millar and L. M. Turk, John Wiley & Sons, Inc.) and similar ones can be found everywhere in scientific agricultural literature nowadays. Nor is this the exclusive knowledge of a small group any longer, but it is general knowledge. If a scientist or agricultural writer were to raise his voice against the importance of organic matter he would be denying the progress of his own science. A further example is by Firman E. Bear in *Soils And Fertilizers* : 'Any system of soil management which increases the soil's content of organic matter, proportionally increases its microbial population, with all the advantages that accrue as a result of their activities. This is one important reason for having a cropping system which maintains a high content of organic matter in the soil...The total microbial population will be closely related to the soil's content of organic matter since this constitutes the primary food of most soil microbes.

'After these organic materials have undergone decomposition to the point at which they may be classed as humus, they constitute an important portion of the colloidal matter of soils. In sandy soils this is of particular importance in that the humus serves in the same capacity as clay to form a binding material for the coarse particles. In clay soils, the colloidal humus tends to bind the finely divided particles together in the form of granules. During periods of drought, these organic colloids are dehydrated, and rehydration is very slow. Their granulating effect is more permanent than is that produced by inorganic colloids.'

One important point in Firman Bear's statement should be impressed upon every organic enthusiast's mind, namely that 'after organic materials have undergone decomposition to the point at which they may be classed as humus...' Much harm has been done by the application of raw, insufficiently decomposed organic matter which ties down nitrogen and bacterial life during the decomposition period. Many an organic farmer has suffered these consequences through ignorance of the fact that science long ago established the conditions under which the application of organic matter as humus can be beneficial. It is of extreme importance that completely decomposed organic matter be used for fertilizer. Here both the opposing camps have made mistakes. The one in recommending plowing or disking under fresh manure as it comes from the barn. To do this leads first of all to losses of nitrogen and ties down soil life until the manure is decomposed or digested in the soil. Hence the observation of the slow action of manure which becomes apparent only after several months or even only in the second crop. Should the first crop be one which needs much nitrogen at once the readily available nitrogen fertilizer gets the upper hand and makes a better showing. It has, however, been observed that the high nitrogen fertilizing produced only temporary and not lasting results, had to be excessive (and expensive) which in turn had unfavorable influences on the physiology of the growing plants (too fast a growth, lots of foliage but reduced seed production, increased susceptibility to pests and diseases, lowered vitamin content). The increase in bulk yield of green plant masses is not always a success. In the case of pasture grasses it may cause bloat in cattle. In legumes it reduces the formation and action of nitrogen-fixing bacteria which then consume instead of fixing nitrogen.

The organic school on the other side, has recommended sheet composting, i.e. the spreading and disking or plowing under of raw organic materials, garbage, etc., which has the same result of tying down the soil life. Many who have tried it have reported to this writer that it was a sure way to spread diseases. Legumes, especially beans grown on land treated in this way showed yellow leaves and yielded poorly.

Firman Bear further says in *Soils And Fertilizers*, page 73-74: *The Effect Of Organic Matter*, 'The application of liberal amounts of manure and the incorporation of plant residues, including the roots, apparently have the effect of developing lines of weakness which prevent the forces of cohesion from being sufficiently effective to permit the formation of clods. Repeated heavy applications of manure may so affect the physical characteristics of a clay soil as to make it take on the qualities of a loam. Occasionally, quite the opposite effect is noted. In soils that are somewhat wet and in which the oxidation of the added organic matter is relatively slow, the application of manure delays the drying-out process, with the result that in early spring it may be difficult to find a favorable moisture condition at which to work the soil.'

Bear is perfectly correct in these last statements. However, the use of completely decomposed, that is composted, manure will entirely overcome this difficulty. The bacterial treatment of manure, compost, and the soil itself, as reported in an earlier number of this magazine, has completely eliminated the dangers mentioned above. When even famous authorities in the field of bacteriology pretend that the organic matter rots anyhow, that bacterial or other preparations are not needed or are 'mystical,' they reveal, regrettably enough, that they have never tested it out and that they are not up on recent developments. The critical designation 'mystical preparation' is unjustified. Something is mystical only when one doesn't know about it.

Interesting data on the results of mulching in reduced water losses is also reported by Professor Bear:

*Tons Of Water Evaporated From Acre Of Soil
With And Without Mulch*
(Data for a period of 100 days.)
Table 38

Depth of Mulch	Clay Loam	Black Marsh	Sandy Loam
None	2,414	588	741
1 inch	1,260	355	373
2 inches	979	270	339
3 inches	889	256	287
4 inches	883	252	315

'Under the conditions of the above test, the losses of water by surface evaporation were practically cut in half by the use of a mulch. 'Mulching decreases or inhibits the capillary flow, and diffusion through the mulch is practically negligible. This practice is very effective in conserving soil moisture and is founded on sound scientific principles. An especially interesting illustration is brought out in the comparison of the loss of water from a soil under arid and humid conditions respectively. As might be expected, the loss at first is much more rapid under arid conditions, so rapid in fact as to tax the soil's ability to move water from within to the surface by capillarity, and in consequence a dry layer is formed which keeps the subsequent losses far below those which take place from the soil under humid conditions, where the capillary flow to the surface persists until the moisture content of the whole soil is very low indeed.'

If all this which outstanding scientists themselves tell us about organic matter--and the farmer knows from experience how true it is-- that he is inspired to build up the humus content and microlife of his soil should be understandable. Any argument against the building up of organic matter in soil is contrary to the interest of the farmer which is to improve and maintain the staff of his life--namely, a living soil which holds or stands up for centuries, which is the basis of progress for any nation in any epoch.

Some critics do not deny the validity of this. However, they do not believe that it can be done. There is not enough organic material in the world by way of manure and compost, even including green manures, to provide the soil with all its needs, they say. This definitely is not an argument against the organic idea. It is rather a challenge. That a fellow is poor is certainly not an argument for him not to earn money or to stay out of business. It is a stimulus to do something, to get going. The average organic matter content of the soils of the United States today is 1.5%. Virgin soils ranged from 4 to 6%. Truck farms and grain farms frequently have soils with a content below 1.5%, yes even below 1% organic matter. The borderline, according to this writer's observation, is about 1.5%. Below this, soils break down rapidly and are unable to 'hold'. They have lost their reserves and the application of fertilizers will not feed and build up the soil but only provide it with the absolute needs for plant growth. The situation is the same as that of the fellow who makes no profit in his business and can spend only as much as he is barely able to make. Many farmers are in this situation today with regard to their soils.

The collection of wastes and their processing is imperative, if we ever want to solve the problem. If our soils continue to decline, the farmer will not even be able to buy fertilizer in the future. To discredit sound humus practices and the processing of wastes in large plants as has recently begun, is about the most shortsighted policy of some scientists and agricultural writers ever to be offered to the farmer.

When mineral deficiencies were discovered everybody jumped on the bandwagon and tried to develop the fertilizer business. Now we see the damage done by humus deficiencies--and many workers in the field are discredited because they try to heal and amend them. The fertilizer boys will one day--we hope not too late for them--discover that increased humus production in soils will make their fertilizer last longer and be more effective if used with reason and will profit from the situation. They certainly cannot help but profit if the farmer has built up reserves in his soils and become more stable in this business. The bank or insurance company with a large reserve capital is considered to be the safest. Equity funds mean gambling or taking a chance. Why not apply the same reasoning to soils?

There is evidence that soils can be handled successfully with increased organic matter, even without additional fertilizers. This writer has grown a wheat for twenty years on organic soils, using the bio-dynamic method by the way, which always yielded above the average of the area with some of the highest yields ever obtained. This wheat also showed one of highest mineral contents, including trace minerals, and a high protein content above the level of its kind anywhere else.

It must, however, also be stated that small amounts of fertilizer combined with the organic method have done no harm to soils. Excessive amounts of fertilizers have reduced vitamin and protein contents. With the proper organic treatment, fertilizers can be reduced to a harmless minimum. The organic movement, nevertheless would be wise to be tolerant. There are many areas where there is no livestock, therefore no manure, where there are no other sources of organic materials. These areas have to continue to use fertilizers until such time as sufficient organic fertilizers are produced to take care of their needs too. It is quite a different problem to work organically in small gardens and fields from that which arises in areas where thousands of acres are intensively cultivated with no access to organic matter. In such regions the use of green manuring is imperative and the only source for the time being. But then, the green manure should be broken down with bacterial action in order to avoid the dangers mentioned earlier. In due time the use of composted organic wastes will develop and improve the situation.

One observation should give food for thought to the most ardent opponent of the organic method. The outstanding property which all soils that have lasted for hundreds of years under cultivation and remained productive have in common is their high content of organic matter and microlife. The soils which broke down, showed erosion, and needed more and more fertilizer but still did not improve, had lost their organic matter and had a low microlife. To simply state that fertilizers have increased yields is correct, but still not the whole truth. The question is 'at what expense have these yields been obtained.' Have we not lost many soils in this process? Have we not lost capital, i.e. lasting soil fertility in spite of increased yields? The fact that soil conservation has become a must should bear out our point.(9)

The facts about soil life are known. Why not let them work toward starting a new, progressive phase in agriculture, creating new industries and processes, as they are already on the march?

Bibliography, Quotations And Comment

1. J. S. Jaffe. The Relative Merits of Inorganic and Organic Sources of Plant Nutrients. *Better Crops with Plant Food*, March 1952.
Chemicals-Fanaticism and Fact. *California Farmer.* May 17, 1952.
2. C. K. Beeson, Plant, Soil and Nutrition Laboratory, U. S. Dept of Agric. pp. 64-67.
Chemicals in Food Products. *Congressional Hearings.* H. Res. 74. Part I. Committee to investigate the use of chemicals in Food Products. 82nd Congress.
This hearing is of the greatest importance, providing reports and evidence not only on the use of chemicals in agriculture and in foods, but especially discussing at large the much disputed use of insecticides. It appears from reading the material presented that a proper labeling law for organic products is as necessary as were the labeling laws for foods, pharmaceuticals and fertilizers in the past. The customer needs assurance that he is getting agricultural products unadulterated, with none or the harmless minimum of insecticide residues.

3. Some of these articles in favor of inorganic fertilizers are probably well meant but they lack careful study of the organic problem, voicing only opinions and revealing insufficient knowledge of actual organic practice. In the scientific literature about organic matter in soils one can find almost all the arguments in favor of organics and one need not resort to the organic literature to contradict these statements.

4. Many organic composting enthusiasts overlook the fact that their practices can easily be followed on small, diversified farms, in small gardens, and in places where organic source materials and manure are readily available. If one deals with large farms, with areas of horticulture, truck farming, monocultures (cotton, citrus fruits, for instance) where there is no organic source material, or if one has to handle large amounts of wastes, municipal garbage for one, at the rate of hundreds of tons per day, then organic methods require adjustment. The mechanical handling of such bulky materials at an economic level is something which needs careful study and experience. It is not a matter for amateurs. The best brains of the country have to study the problem and really create organic engineering as well as an organic science in general.

5. One need only scan the scientific literature and agricultural textbooks of the last two years to find about everything in favor of organic matter and its importance. The very same literature which contains data in favor of minerals also reports plenty of data substantiating the need of organic matter. One of the best brochures on the subject was, for instance, published by the National Fertilizer Association. Pamphlet No. 136, *Organic Matter the Life of the Soil.*

Also interesting with regard to obtaining top results with organic matter are: *Studies with Organic Materials for Vegetable Crops* by L. M. Ware, W. A. Johnson. Agricultural Experiment Station, Alabama Polytechnic Institute. Bulletin 280, June 1951.*The Wrong Turn* by Firman E. Bear (Rutgers University). Journal of Soil and Water Conservation, Vol. 6, No. 2. April 1951.

'....Usually a nitrogen top-dressing or side-dressing produces such marked growth and color effects as to be very readily seen, assuming that a check area is available for comparison. This means that the supply of organic matter, the storage agent for nitrogen, is inadequate. Other evidence of lack of organic matter is found in the increasing tendency toward minor-element deficiencies in cultivated soils that have been brought to a high level of productivity by heavy use of concentrated inorganic fertilizers.

'Instead of looking with a jaundiced eye at the efforts of the organic-farming enthusiasts to develop organic matter supplies for use on the soil, the fertilizer industry would do well to interest itself in this problem. There is need for study of the possibilities for recovery of city wastes....

'These data point in the direction of the need not only for more organic matter, but a variety of kinds of it, no matter whether we are thinking of a cover-crop system or one involving sod crops. Many weeds make highly important contributions in mobilizing minor elements in the soil. Ragweeds and lamb's quarters, for example, are excellent accumulators of zinc. It is conceivable that they might be deliberately grown for the purpose of mobilizing this element.

'Sod and cover crops should be liberally fertilized with complete fertilizer. Nitrogen alone is not desirable, since this tends to develop top growth at the expense of roots. A complete fertilizer helps them to develop not only larger tops but larger root systems as well that accumulate more mineral elements from the lower soil depths. When such sod and cover crops are worked into the soil, they provide excellent food for soil microbes, and these microbes liberate large amounts of all the essential mineral elements, including nitrogen, from them.

'The fertilizer industry represents the most important chemical development, in terms of human values, the world has ever known. Fertilizers can be made to stand between us and any possible deficiency of food for centuries to come. But fertilizers alone, no matter how heavy the rate of application, will not meet the requirements for soils that are producing cultivated crops. Soil must be fed organic matter in larger amounts than the roots and residues of such crops can provide. This requires manure and well-chosen mixtures of sod and cover crops. If, in addition, we can develop means of producing supplemental compost from city refuse, or by any other means, so much the better. In my opinion we should lend assistance to efforts that are designed to avoid waste of organic materials. It might well pay to subsidize the processing of such organic wastes as can be made available for use on intensively cropped land.'

Textbooks like the following contain valuable material for the appreciation and understanding of organic matter in soils and microbial action: *Soils and Fertilizers* by Firman E. Bear, John Wiley & Sons, *Fundamentals of Soil Science* by C E. Millar and L. M. Turk. John Wiley & Sons, Inc. New York.

6. *Soil Organisms - Fact and Fiction* by Dr. James P. Martin and Jarel O. Ervin. Western Grower and Shipper. May 1952, Vol. 23, No. 6.

'In most soils there is insufficient decomposing organic matter to meet the nutrient requirements of cultivated plants during the period of maximum growth. Even if organic manures are added, it is not certain that sufficient nutrients would be liberated to give optimum plant growth. Certainly if all farmers were forced to rely on organic fertilizers only, the available supply would not begin to furnish the required amount for good crop yields. Our high production is dependent on substituting or supplementing organic with inorganic fertilizers. Further increases in crop production will result largely from increased use of inorganic fertilizer and better soil management practices which include the use of green manure crops, rotations, and organic manures when available.'

If green manures and organic manures are part of the better soil management, so this writer thinks, why not improve the effects of green manuring, manures and composts as outlined elsewhere in this issue? Why forget those who are trying to improve the situation and why not try to get more organic fertilizers if there are not enough. The above quoted scientists report in a table bacteria counts of soils, untreated and treated with cover crops and calcium nitrate fertilizers. The highest count of bacteria they report is 35 million per gram of soil. In our practice on skillfully treated organic soils, counts of 500 to 800 million per gram are not rare.

In talking with hundreds of farmers and growers all over the country there is one phrase one hears over and over again: 'We know we can grow high yields with fertilizers but we also know that we don't do any good to our soils with them. We have to learn now how to build up our soils.' The experimental stations would do well to heed this remark and work with the farmers and growers. At the present moment there is a deep cleft between the scientist, the farmer and the fertilizer man. The promise of higher yields is frequently based on an erroneous concept. Yields cannot be increased indefinitely and not on a constantly rising curve. The limiting factor is water. If there is a lack of water or an insufficient supply of it, no fertilizer will increase the yield any further than the moisture reserves of the soil will allow. Here the importance of organic matter becomes clearer than anywhere else, for it increases the water-holding capacity of the soil.

7. *The Inorganic Side Of Life*, by Firman E. Bear. Better Crops with Plant Food. April 1952.

'Considered in relation to the needs of animals and man, two points should be emphasized. First, by using excessive amounts of potassium fertilizers the potassium content of plants can be raised to such levels as to materially reduce the calcium, magnesium and sodium contents. This is contrary to the best interests of the consumers of these plants, since from the quantitative point of view calcium is much more important than potassium to animals. Secondly, in proportion as acre yields are stepped up through the production of more carbohydrates, as in the case of hybrid corn, the percentages of all the minerals in plants are lowered. Potassium is known to be of value in the animal body, but an excess should be avoided...gross tetany is related to too high a potassium content...as a result, the amounts of calcium and magnesium in these plants are greatly reduced. The imbalance between potassium and calcium interferes with the normal rhythm of the heart. Similarly, a lack of balance between calcium and magnesium affects muscular activity...the magnesium content of the blood of these animals is often very low.

'Much more attention is being paid to minor elements now than formerly because of widespread deficiencies. Some of these deficiencies are a result of serious soil erosion, with consequent loss of organic matter. Others are due to losses of the elements by way of harvested crops during many years of farming. The substitution of the tractor for the horse has added to the difficulty, since this often means more acres under cultivation and less manure for use on the land. Higher acre yields, obtained by planting hybrid seed and using larger amounts of NPK fertilizers, have increased the need for these elements in the soil. And the ever greater purification of materials that go into the manufacture of fertilizers has added the final touch to the trouble.'

Prof. Firman Bear, an authority on fertilizers, should be commended for the objective stand he takes in these matters as well as for his appreciation of organic matter in soils, the use of organic wastes, the importance of weeds as sources of trace minerals.

The mere addition of the missing trace minerals to the fertilizer or soil doesn't solve the problem for these minor elements need to be present in a form which avoids the locking up in the soil but renders them acceptable to the selective action of plant roots. For the record, it may be mentioned that this writer more than a decade ago already pointed out these problems, for instance the potassium problem in his book *Bio-Dynamic Farming and Gardening.*

8. The importance of microorganisms in soil is being recognized more and more widely. It is a peculiar phenomenon that an authority in soil microbiology who has actually laid the foundation for the understanding of this microlife, such as S. Waksman objects to organic practices which are a direct consequence of his own findings. But then, we have the fact anyhow that, for decades, applied soil microbiology has been treated like Cinderella in agriculture; most of the researchers and practitioners being fascinated by Liebig's mineral fertilizer theory. Fundamental works on soil microbiology have been written as far back as the beginning of the century. But the knowledge contained in them has never been practically applied until very recently. Had our agriculturists carefully studied Liebig's own work in all its details they would have discovered that towards the end of his career he had certain doubts in regard to his own mineral theory. He discovered that certain soils did not fit into the theory. These were heavy, dark brown humus soils in the Danube plains which yielded high in spite of 'apparently' no fertilizer being applied. Also the testing of these soils did not reveal the true situation of their fertility. These soils were high in organic matter. It would appear that the Liebig theory applies to worn out, mineralized soils, but not to soils with a high organic rating. Just imagine, there would have been no soil erosion, no need for soil conservation, no trace mineral deficiencies, if Liebig's doubts had been noted and taken as seriously as his fertilizer theory. It just demonstrates how a one-sided

point of view can produce an apparent success for a short time followed by a bitter awakening.

An excellent little brochure on the subject is *Microorganisms and their Effects on Crops and Soils* by T. M. McCalla and T. H. Gooding. Nebraska Agricultural Experiment Station. Circular 90. March 1952.

9. The newer concept in soil fertility and microbial activity is probably best described in an article by F. Lyle Wynd, *Feed the Soil,* The Scientific Monthly, Vol LXXIV, No. 4, April 1952, page 223. This is undoubtedly one of the most important publications of the year. Men like Firman Bear and Lyle Wynd merit commendation, for they have grown up and become renowned scientists with the fertilizer theory. But they have not stood still with it and are ever searching scientists on the path to a deeper understanding of soil life. The following quotations from Lyle Wynd's article more than substantiate what this writer wishes to express.

'There is an ever widening rift between the policies of the fertilizer industry and those based on the results of research in the fundamental properties of the soil. This rift can no longer be ignored by the fertilizer industry for already it has lessened significantly the effectiveness of the industry as a partner in the general agricultural program. Somewhere along the line of its industrial program the fertilizer industry has missed the boat.' He then mentions the two schools, the need for cooperation. 'This cooperative effort,' he continues, 'depends on the well-placed confidence of the farmer in the organizations that sell him fertilizer, and such confidence is incompatible with the present confusion and uncertainty. What the farmer does not realize is that he is the victim of an error and that he has fallen into the seductive arms of modern advertising. The error is sometimes forced upon him in ignorance, but many examples of advertising propaganda could be collected that are so obviously absurd that one is forced to believe that the farmer is the intended victim of a deliberate campaign.'

These are the words, not of an organic faddist, but of a professor at Michigan State College. One can only hope that he can stay there and will not be out of a job for these courageous words.

He demonstrates that fertilizer is not plant food at all (the favored term of fertilizer legislation and formula declaration) but at its best can be only food for the soil; that plant food is contained in the humus of soil and prepared by microorganisms, i.e. made available by them so that roots can absorb it. 'We believe that the concept of fertilizer as a specific plant food is not entirely correct, and that the misconception has been popularized by the advertising program of the fertilizer industry. We have no quarrel with the industry itself. Our only intention is to call attention to an error that has resulted in so much harm to the industry and to the farmer....Soil scientists and plant physiologists, however will tell the farmer that plant foods are components of the colloidal and dynamic biological complex of the soil, and that plants obtain their foods from complex organization, in a manner of speaking, only by permission of the soil.

'Our discussion will be based on two fundamental characteristics of soils. The first is that soil is biologically alive and second that its colloidal properties govern the release of food to the plant. We might say, in other words, that the nutritional aspects of soil fertility depend on activities of living microorganisms and on the electrical properties of its non-living, colloidal components. Our first task will be the consideration of the organic life in the soil, because all that the soil is or may become depends on it. Soil is alive, truly, actually, and literally alive. Soil is not soil until it is alive and no amount of chemical plant food mixed with dead, finely ground rock particles would produce the equivalent of a productive soil.' (Again, for the record, it may be pointed out that this has been the main thesis of this reporter for over two decades.)

'Soil must be something more than a mixture of fine particles and chemical plant food. This 'something' is the result of the living nature of soil and it must be the basis of permanent soil fertility. It must be preserved. If it be preserved, all is well. If it be not preserved, no amount of 'plant food' dispensed from sacks will itself remedy the loss.'

Wynd emphasizes the need of living organic matter, teeming with microorganisms. 'A soil was never fertile because it contained inorganic plant food and dead organic matter. If we admit that the living microorganisms in soil must thrive in order to maintain and augment fertility, then we must also admit that the maintenance of this living population is of great importance. These organisms require food...'

'The supply of nitrogen, phosphorus, and sulfur, and of many other plant foods as well, is completely dependent on the metabolic cycles undergone by the micro-organisms. If we think of fertility in the long-time sense, we must first think of the biological and colloidal requirements of the soil itself and only secondarily of the nutritional needs of specific crops....It is amazing to a soil scientist, and he chuckles inwardly with pedantic satisfaction, when he hears a farmer's lament that he has added so many pounds of phosphate per acre and obtained no significant increase in the yield of corn. It is lucky for all of us that he was unavoidably feeding bacteria and thereby preserving some of the essential properties of soil for a future and more intelligent master...'

'Any concept of fertilizer as plant food, to the exclusion of its importance as a soil food is based on the belief that the plant can be fed directly without first feeding the soil. Such a niggardly practice is uneconomical for the farmer in the long run. His soils will unavoidably deteriorate if he practices it over any considerable period of time, and this will force him to use more plant food, with still greater deterioration of the soil. Such a practice is contrary to a sound agriculture, contrary to the ideals of the conservation of fertility, and therefore antisocial and unpatriotic. The long continued use of mineral fertilizers as a plant rather than as a soil food always leads ultimately to a decline in productivity. Because this decline is extreme in some instances, there is an increasing resentment toward the use of mineral fertilizers among a group of fanatic faddists. Unfortunately for the fertilizer industry, the argument of these fanatics is based on facts making their absurd conclusions embarrassingly difficult to attack.' Let's leave out the fanatic faddist and get down to brass tacks, in this case the maintenance of soil fertility, humus, and microlife and there will be sound cooperation.

A COSMIC ERROR

The earthworm spends a life of toil
Producing rich and fertile soil -
All entomologists agree
He does no harm to plant or tree -
But gives his efforts to enhance
The value of their sustenance.

The chinch bug, moth, the weevil and
Some other insects all depend
Upon the products of the wheat;
It isn't flour to them, - it's meat.
They do far more than storms or frosts
To elevate production costs.

And yet a cruel trick of fate
Makes earthworms valuable for bait,
And, owing to the fisherman,
Their lives are but the briefest span;
But milling pests in loot engage
Until they perish of old age.

The reason why, I do not know,
That matters should be thus and so -
But, dug up for a fish's meal,
The earthworm gets a dirty deal,
While in the wheat field or the mill
The hungry beetle eats his fill.

> Harvy E. Yantis, in the Northwestern
> Miller of a Quarter of a Century ago.

IV
THE DECLINE OF AMERICAN AGRICULTURE

According to the 'law of the minimum', the mineral nutriment present in the soil in minimum amount determines a plant's growth. If, for instance a plant is abundantly provided with such essential chemicals as potassium, nitrogen and calcium salts, but lacks phosphates, it will suffer in proportion to its phosphate deficiency. The discovery and application of this law figure prominently in the history of chemical agriculture.

But the concept of this law is being biologically extended. It is becoming a 'life' concept. First, progressive agriculturists grew aware of chemical deficiencies; then of deficiencies in such rare metals as zinc and copper; but we still had poor crops, erosion and dust bowls, for now it was the soil-and moisture-holding humus which was 'present in minimum'. Whereupon more attention was given to a proper, soil-conserving crop rotation, with especial emphasis on leguminous plants. But still the concept was not wide enough. Something was being neglected: man.

In agriculture - since it is a complete system - the 'law of the minimum' covers every life-fostering factor; and man is a life-fostering factor. Indeed, his contribution to the process of growth is enormous. Left alone, nature would try to create a balance of vegetation in accordance with outer circumstances such as climate and soil, or, if the situation was hopeless, abandon it to desert; but would never grow 'crops'. Man grows them for food and raw materials. He owes his existence to partial separation from nature and lives as he pleases. But the forcing of nature in crop-growing disturbs nature's balance, and requires the taking of certain steps to restore the soil to health. Thus man's understanding or not understanding and cooperating or not cooperating with nature come under the 'law of the minimum'; and, while considering the chemical interrelations between light, warmth and the heredity and spontaneous mutations of seeds, we cannot with impunity forget the most important factor of all: man himself. If one factor in a vast complex, we have said, is 'present in minimum', the whole will be thrown out of balance. Today it is man who is 'present in minimum' - and we are beginning to reap the consequences: duststorms, floods, soil erosion, sickly crops, rapid desert formation, degeneration of plants and an increase in pests.

In this agricultural tragedy, America plays an important role. For it was America which brought to agriculture an extreme mechanization. Compare the humanity-crowded, hand-hoed fields of the Near and Far East with the 'empty' tractor-worked fields of the average American farm, where one man is made responsible for a hundred or more acres. 'The machine has replaced man', we say - and strain toward that time when all hand labor which can be assigned to machines is so assigned; when man is little more than the servant of the machines he invented.

Is this good - this suppression of man's creative function in the life of nature?

Luther Burbank, the California plant wizard who has created many new plant forms, said one day: 'It is finally I, the human being himself, who gives birth to plant life, who puts in the seed, changes the forms. Man is the determining factor who gives life or takes it away.'

Man's destiny is to become the master of nature; something higher than his present role of pupil and experimenter; to use nature as a tool.

But apprenticeships are beset with difficulties; major or minor catastrophes which, coming at the end of every period, and provoking not only pain but a deeper understanding, invite a forward step. If the apprentice ignores any factor in the complex of his work, he may fail of his aimed-at mastership - and today, because of man's overemphasis upon machines, and increasing underestimation of that factor which is himself, he and the whole of agriculture face a grave danger.

For many things on a farm cannot be done by machines, such as feeding, looking after animals, stable-cleaning, weeding, combatting pests, and milking.

The latter furnishes an especially good example. A person who has never learned hand-milking can assist at a mechanical milking, but does not understand the finishing off by hand, the all-important 'milking out' - the very process which determines the health and production of the cow.

Present-day farm labor is poor in both quantity and quality. The situation has been aggravated by the draft, and by the disproportionately high prices paid for workers in defense factories, as well as by the widespread notion that skilled farm labor can without hurt be replaced by the unskilled - even boys and girls of teen age. The younger generation of helpers, machine-minded and machine educated, feel embarassed when asked to do hand work and often refuses; with the result that they handle inexpertly such tools and implements as the scythe and sickle, the team-dragged plow and the sharpener for the blades of a mowing machine - as if driving a tractor only were identical with farming.

But perhaps most of the blame for this exodus from the land should be laid on a psychological factor. Man has lost or is losing his appreciation for nature. Taking it for granted that he will always be provided with an abundance of everything, and liking the spectacle of others working for him, he disassociates himself, mentally as well as physically, or physically because mentally, from the land - settling down, satisfied, in man-made, technical surroundings and to an artificially constructed life. Monotonous city sidewalks, factories with rattling machines, a crowded downtown office lit almost all day by electric lights: any place for work seems preferable to the always fresh air and wide horizons of a farm. Nature is in a kind of Cinderella position. Thus do values get all mixed up in the mind of man.

What will happen within a few years after the older, landskilled generation has died out? This is a serious question, troubled over by many a conscientious farmer. Industrialization and city life teach little farm lore. Refinements of soil treatment and 'complicated' soil-building crops may be progressively abandoned, with only a few 'mechanized' crops surviving. Before extreme standardization, colorful diversified farming tends to go down, and one-sided crops, as monotonous to the soil as to the eye, exhaust the soil's fertility. Anticipating this, the farmer of 1943 is developing a 'Goetterdaemmerung' mood. He fears he is getting a pre-taste of dreadful things to come.

About the plight of the farms the city man worries less. Yet he has plenty of cause to worry, for does not history offer conclusive proof that the life-blood of a nation derives from its rural population? This is particularly true of a 'young' country like America, where most of our great-grandfathers and grandfathers were still connected with the land. In Europe many centuries old families have survived largely because of their unbroken contact with the earth, that source of regeneration. Statistics, as published by O. E. Baker of the Department of Agriculture, demonstrate that it is the rural rather than city populations which lead in replacing a nation's losses through death. Baker mentions the period between 1950 and 1960 as a kind of climax, the time when the death rate will have outdistanced the birth rate, in the United States, and the population will begin to decline.

The fate of France is a terrible object lesson. An increasing urbanization, together with an idea of 'security' which ignored that greatest security of all, a self-sustaining farm, created an urban proletariat; and, for lack of proper help, more and more farms had to be abandoned. Future historians will see in these matters some of the reasons for the fall of France.

Just before the outbreak of the Second World War, the author discussed bio-dynamic farming and gardening methods with the Director of the French Ministry of Agriculture, emphasizing particularly the necessity of maintaining the land's fertility by means of a long-range agricultural program. The conversation ended thus:

Mr. B.: 'You talk, Dr. Pfeiffer, like someone who wants to establish a dynasty of peasants. But what are we to do today with those who want to make a profit, and for whose profit we are responsible?'

Pfeiffer: 'If a long-range agricultural program on a bio-dynamic basis is not set up, the fertility of the land and the health resources of humanity will suffer. Thus the possibility of a renaissance of the French nation depends upon a dynasty of peasants coming into existence. If one worked only for monetary gain, one would live at the cost of the capital of the earth's natural fertility. The nation would then be of short duration. It is part of the Minister's responsibility to determine whether the short-term gain will cheat the country in this regard, and what measures, if any, should be taken to insure durability for the coming centuries.'

Two years later, in the wake of unspeakable catastrophe and suffering, we read:

'Vichy, July 3, 1940. Guiding Principles of the French Labor Ministry for Reconstruction: a. France is above all a nation of peasants and artisans. These occupations, all too long neglected, must revive. A sensible balance between peasant and industrial economics must and shall be found. b. All non-specialized workers whom war drew away from the land must be brought back to the land. c. A general policy of repopulating the land must be taken up. French soil can give employment to and feed many more people than in recent years.'

What immense suffering could have been avoided if France had followed up such policy fifty years earlier! If the French had remembered that farming is still the 'profession royal', in that it protects the water-supplies, provides the human being with almost all the necessities of life, and maintains (and this is perhaps as important as our bodies' welfare) the surface of the earth in a state of fertility consonant with adequate surroundings for human culture! What does it all add up to?

The fact that agriculture all over the world, but especially in America, is threatened by death, is in a dying state, but can by proper treatment be raised up again in health. The 'treatment' in this case is a more fruitful philosophy of the land - one which reverses the wheel of retrogression.

Unfortunately the human being has developed a psychological peculiarity. Neglecting necessities, he prefers theories, ideologies, trivial hobbies, even illusions - as if these are the realities, and necessities only the burdens from which to escape. History has been a series of escapes and returns.

But there is a comforting thought. It is not yet too late; and the war years are helping us to see the gravity of the situation - to see how quick is the drop from abundance to shortage.

The whole problem is primarily a moral one. Our future depends upon our choice between death forces and life forces; upon whether or not we will return in humility to the soil. The great questions are: Will we return to a philosophy of life which lays stress upon growth? Will our youth be educated in the spirit of growing things, and of service to life? Will they learn that it means more than money to plant our seeds and harvest our crops? If the right inner attitude towards the soil penetrates the human race again, a renaissance of rural life will begin, and not only will new resources be created for our population, but spiritually we will be 'safe'.

Surely our youth, returning from the horror of destroying (for is it a pleasure, a lust?), will appreciate the spirit of allowing things to grow, of building up, which has sunk to a 'minimum'.

The desire to rebalance Man and Nature should be in our souls a hunger and a thirst.

V
HOW TO BUY A FARM

More and more people are thinking of investing their funds in agricultural pursuits. It is, therefore, timely to give some definite advice on this question. Let us then first discuss the motives of these prospective buyers. Quite a number of them evidently look for an escape from the inevitable further loss in the purchasing value of the dollar and they intend to pick up land to rent it out for sharecropping or cash. The man who actually works the land will thus have not only to make a living for himself and for those dependent on him, but also to pay someone for the privilege of so doing. Few improvements can be expected under such conditions. On the contrary, the tendency to industrialize agriculture develops mono-cultures and forced production appears with the exhaustion of the soil, erosion, dustbowls and final destruction of the organic living soil, humanity's most valuable inheritance. The following advice is, therefore, intended only for present or future farmers or people in close contact with farming. We believe that it may bring some help even to those of vast experience. If you have practically no experience as a farmer, go out and get it for a year or two before you settle down in the country. By all means buy your farm meanwhile if you are convinced that the right opportunity has come your way, but do not do it unless you have had absolutely competent and reliable advice. Be careful, however, not to start on too large a scale for your cash resources. You can almost always enlarge your place later on if need be.

Buying a farm is quite an art and requires much wisdom, since there are many questions involved besides the price of the land, problems which are often overlooked when one first sees the place may become a burden or render impossible the entire project. We suppose that you want to make a living out of your farming. Forethought must, therefore, be your watchword. In selecting the proper site be led by reason rather than by sentiment. You will find troubles enough without the handicap of having to make an overcapitalized place pay its way.

The first question to consider is the QUALITY of the SOIL, not only its actual condition, but also its original, virgin producing capacity to which it may be restored if to some degree exhausted. This degree of exhaustion must be explored, of course. The soil may be in part or wholly run-down so that there is little chance to restore it to health. You must find out how many years it will take to effect necessary improvements and what the maximum production will then amount to. Otherwise, you cannot know the proper basis for your investment.

The next matter to look into is the MARKET and TRANSPORTATION situation which should be particularly well studied if you intend to grow specialities for sale, such as vegetables, fruit, etc. Then you must consider the LABOR SITUATION in all its details, the quality of the help available in the particular region, steadiness of employment there, wages, etc. If you need a farm manager, look for a reliable man with initiative, experience and intelligence. Near towns or industrial areas you will find less skilled and more 'movable' labor; far away in the country you may have more difficulty to pick a man, but he may have better standards of work.

CLIMATE and WATER CONDITIONS are of great importance. An ample supply of good drinking water for man and animals is essential. You should find out all about the rainfall during the different periods of the year. The yearly rainfall may appear sufficient, but if there is regular drought in spring and early summer, you will find yourself up against it for good pastures, meadows and clover fields. You should also pay attention to WINDS and available protection against them. Otherwise you may have to put in unusual crop rotation, hedgerows and additional cover crops instead of cash crops.

The state of the LIVING QUARTERS, stable, barn and other dependencies must naturally have keen attention in order that you may judge the amount of repair needed. Do not forget details, such as repairs to gutters, drainage, roofing, plastering, new window-sills, painting, heating, light installations, road repairs, a deep well, etc.

You should find out about the state of HEALTH of the LIVESTOCK belonging to the man who is turning over the farm to you. If he has had trouble, you are liable to inherit it. If you still want the place, you would be wise to start with a cheap but healthy type of cattle.

Do not forget that a farm that has been abandoned or badly neglected will require considerable extra time and expense to be cleared of weeds and harmful insects. It also will require a longer period for soil improvement.

Finally, make a preliminary check-up on your available capital in relation to the particular farm, the purchase of which you are considering. There is the money needed for the purchase itself; then money needed for repairs, etc. of buildings; money for roads, fences, hedgerows; for machinery, tools and materials; money for seeds, livestock and, to start with, for feed and perhaps for fertilizer. Then you must have funds for the first periods of wages, taxes, and general running expenses, until the income shows up. A good way is to use one-third of the available capital for purchase of the property, one-third for livestock and equipment and one-third for expenses and losses during the first years until things are running smoothly. If you can follow this golden rule, you should meet with ultimate success, but the further you deviate from it, the nearer you will come to failure. You should not mortgage your farm in order to buy both cattle and equipment. Then you merely work for the benefit of the mortgage holder. The maximum mortgage should, if possible, not exceed 25% of the PRODUCTIVE land value.

Here it may be well to issue a few warning remarks. Do not purchase a 'view' only. Make sure the view carries with it also good soil. On the top of a hill, the good soil is frequently washed out. It is an expensive undertaking to restore such a devastation. It takes years to produce a new fertile top layer. A hilltop should be covered with trees in order to be protected. If that should not be the case, you are entitled to be suspicious as to the fertility of the land.

Also, never buy just because the land is cheap. There is almost always some specific reason for such cheapness that will take a great deal of time, work and money to cure. There is usually something wrong with a farm advertised for sale. The soil may be nearing exhaustion or erosion may be coming on fast; the buildings may need more repairs than the former owner could afford; market conditions may have changed for the worse; cattle diseases may have caused losses, etc. I remember someone who bought land for $10.00 an acre and spent

$150.00 per acre for the clearing of it. Even if the surrounding farms should be excellent, you have no guarantee that the soil you are investigating is a fair buy. The best chance of acquiring a good farm would be when the owner wishes to sell because of illness, old age, etc., or when the property has been inherited by people who have no interest in continuing the farming activity. In such cases you are likely to get your equipment at a reasonable price, too.

When you have taken all these things into careful consideration and still think that you have found a suitable property, it is time to go into further details - but do not be hasty.

For the preparation of the budget, consider the following points:

A. CAPITAL INVESTMENTS

Purchase of the land. It has to be realized that the purchase value does not always indicate the productive value. For some such reasons as nearness to a town or suburban area, as very good farming country, etc., the price paid may be higher than the amount of money on which it could produce an average interest. The difference must be considered a loss unless the farm can be resold at a higher price, but this is speculation and has nothing to do with farming.

Building Repairs and eventually new buildings, road repairs, reforestation, drainage, reclaiming of land, the initial clearing of space, well digging, fencing and hedgerows are investments to be capitalized and added to the purchase price in order to arrive at the true capital investment.

Investment in Farm Machines and Tools may be capitalized except for small tools which wear out fast and or get lost easily. Most of the farm machines should last for 10 years or more, but some wear out in considerably less time. Use good common sense in the depreciation rate.

Investment in Livestock and Other Farm Animals. You have to decide whether you want a dairy or beef farm, a cattle ranch, a goat or sheep farm, a generally diversified farm, or whether you wish to specialize on the growing of some particular crop or crops (for example, 80 acres of celery).

The number of cattle necessary to stock a farm should be determined by the amount of manure needed and the amount of feed available, including hay, pasture and grass. The bio-dynamic method sponsors only the idea of a diversified, self-supporting farm. Once the organization of the farm is complete, all necessary manure and feed should be produced on the place. The calculation of these factors determines the amount of livestock. The local market conditions indicate whether a dairy is more advantageous than a beef herd. If you are a beginner, you should not start with a valuable breeding herd. Healthy grade animals will be to your advantage until you have acquired experience in breeding. And remember that no farmer willingly parts with his best cow, bull or heifer. If he sells some of his animals he has a particular reason. The only exception is the sale of young livestock when the breeder has to dispose of his surplus. But even then he is liable to hold back the best specimens. In other words, home bred cattle are better than purchased cattle. As long as you continue to purchase cattle you cannot permanently improve your herd.

The value of a dairy is that it provides a steady income, but it is a complicated and troublesome business and needs a skilled dairyman. Beef cattle are easier to keep, especially when labor is scarce. However, this business is more speculative since it may take 12, 18 or more months before you have animals ready for sale. Also much experience is needed to judge whether calves may develop properly or will not even pay for their feed.

The type of cattle to be chosen depends entirely on local conditions and no general rules can be given.

The capital invested in livestock may be depreciated or included in the general turnover. We can count on about three years from cow to calf to cow again. Furthermore, not every calf is a heifer, so that probably only half of your initial investment reproduces its value in three years while the other half merely yields a fair interest in the most fortunate cases.

Estimate the number of horses you need and what other animals you would like to have, such as sheep, pigs and poultry. They belong to a diversified farm and should pay for themselves somehow. If you must specialize, do so only if you are an expert. You cannot just buy a hundred goats or a thousand pigs and make money. As soon as you have to buy feed, you are likely to run into a loss, particularly if you are inexperienced.

Size of Farm. Considered from the point of view of needed equipment, there are economical and uneconomical farm sizes. With one man and one tractor, you can till easily 50 acres, but scarcely 100. Such a one-man and one-tractor farm is practical up to 60 or 70 acres but uneconomical for thirty acres and a really hard task for more than 70. A two-man, one-tractor and one-team farm of 125 acres is practical, but uneconomical for only 70 acres because of the expensive equipment which is not fully utilized. With the same farm equipment you can work 30 acres or 150 acres. In the first case, you are proportionately over-invested, unless you make up through specialities which need intensive work on a small acreage (truck gardening for example). You have to decide on the type of your farm, whether it is to be the one-man, two-man or the three-man type, etc. the one-or two-tractor type with one or two teams of horses, mules or oxen, and so on.

SPECIFIC QUESTIONS AND ANSWERS:

Can any farmer introduce the bio-dynamic system? Yes, but only to the extent that farmyard manure and or compost are available in sufficient quantities.

Why is farmyard manure needed for the bio-dynamic method? It contains organic substances derived from the digestive systems of animals. The excrements are rich in protein, cellulose and other organic substances, which in their turn nourish the soil bacteria, earthworms, etc., and through further decomposition provide nutrients for plant roots.

Why could we not just plow under fresh manure and leave it to nature to take care of it? Fresh manure in contact with air and moisture decomposes rapidly and loses 50% or more of its nitrogen, if exposed to rain also up to 20% of the minerals.

Could we not plow under manure deep enough to avoid the oxidation or washing out, covering with a thick enough layer of earth to protect the manure? Yes, we could, and as a matter of fact it is frequently done, but we risk having the manure so well buried that it remains inert and acid without becoming available as nutrients for soil life and plant roots. This is particularly the case if plowing is deeper than 5-6 inches and in heavy clay or very moist soil.

What do you consider the best way of handling manure? Its transformation into humus before it is plowed under. Humus represents the most fertile state of the soil and maintains the soil life. If manure is brought to the soil in the form of humus, it will be absorbed immediately and made available for plant and soil life.

How is this transformation into humus brought about? By proper pilingup of manure and covering it with earth and the insertion of the bio-dynamic preparation or the Bio-Dynamic Compost Starter.

How quickly can a farmer introduce the bio-dynamic method? To the extent that he has manure available. If one has 100 tons of manure per year and uses 10 tons per acre, he can cover 10 acres of land. Provided the manure was rightly treated and handled before plowing under, this farmer has started to convert 10 acres to the bio-dynamic method.

How long will it take then, to convert a farm entirely? As soon as every field has had at least once a full dressing of treated manure it can be considered as converted. If a farm has 8 arable fields and 2 are covered every year with manure, it would take 4 years to convert this farm.

Does this imply then that in four years we should expect to receive the full beneficial value of the new method? Not at all. This is only the beginning. While we lost previously 50% of the nitrogen and much if not all of the humus producing qualities of the manure, we are starting now to stimulate and foster the humus production in the soil itself. This is a life process which can grow and improve only gradually, adding more and more to the soil. A complete conversion can be achieved only when the addition of treated manure is repeated. As a matter of fact, the more frequently this gift of manure is repeated, that is, the shorter the interval, the better the result. It has been our experience that usually the full effect become evident only after the second addition of manure to a given field. This is, however, dependent upon the crop rotation.

So the crop rotation comes into the picture? Does it matter what kind of rotation we have if only we provide manure at certain intervals? Yes, there is quite a difference. We speak of balanced and unbalanced rotations. Since there are soil exhausting, neutral and even humus maintaining crops, the proper rhythm between these three types is essential to preserve the fertility. It would be wrong to plant exhausting crops, such as corn, year after year or to alternate exhausting and neutral crops only. Fields treated in such a way would require an amount and frequency of manure application that would be practically and economically impossible to maintain. A beneficial rotation is one which on good soil has had a manure (i.e. an exhausting) crop, followed by a neutral and then a restoring crop; or on a poor soil, a manure crop followed by a restoring crop and then a neutral crop. In bio-dynamic terms: the conversion of a farm progresses with the crop rotation - for example, (1) corn with manure, (2) winter wheat, (3) oats and clover, (4) clover, (5) soy beans or barley. A farmer with a five year rotation will need five years to convert the entire farm. After the tenth year he will reap the full benefit of the new method. With a four year rotation, he will achieve the goal in a four and eight year period.

VI

SHALL WE PLOW?

It seems that Edward H. Faulkner's book *Plowman's Folly* has turned out to be the best seller in recent agricultural literature. Without doubt, this book stirs up many problems in addition to pointing out one of the most interesting: shall we plow or not? Against the background of the preservation of the organic content in soils, Faulkner develops the idea that plowing is unnecessary and even detrimental to most soils. He emphasizes the importance of humus as fundamental to a soil's persistent fertility. From the point of view of the bio-dynamic method, we find this theory sound and worthwhile investigating. However, we may not need to be always as radical as Faulkner is in his conclusions.

One of the outstanding pioneers in the early bio-dynamic days came to the conclusion that with the soil improvement as accomplished by the bio-dynamic method of farming it would not be necessary to plow anymore. The writer of this article has seen farms, one particularly of several thousand acres in Czechoslovakia, where for many years no plow had been used - the manure being harrowed or disked in. These farms had maintained their organic humus soil with a highly crumbly structure and managed everything with harrow, disk and cultipacker. To him, therefore, Faulkner's book presented a familiar experience. However, there may be situations where the use of a plow cannot be avoided - for instance, in turning grass sods, to turn under a heavy growth of fall weeds, to break a hard surface crust on heavy, moist soil in a humid climate where the disk would not work.

A field roughly finished with all the rubble and stubble lying on or near the surface, may also be undesirable if a field is to be prepared for fine seed such as clover and alfalfa. There a smooth surface is required in order later to have a smooth motion and clean cut at harvest. The writer is glad to report that Mr. Faulkner has called his attention in a letter to 'the mention made in the latter part of Chapter 7 of ways in which the moldboard plow may properly be used to accomplish the desired result. Sod land is one of those places; also stony land and steep hill land.'

Considering all the pros and cons, we may find the truth and the proper practice by investigating the question: why and when does plowing do harm?

In olden times a kind of 'plow' was used with no damage to the soil at all. Soils treated with this 'plow' maintained their fertility for thousands of years, as was the case in Egypt, where, in fact, this 'plow' is still in use on old-fashioned farms today. It is a wedge-shaped, primitive instrument which does not turn the soil but only breaks the crust and hardpan (if there is one), and stirs up the soil. Trouble started only when this plow was replaced by the iron or steel implement with the moldboard, e.g. with the invention of the plow which 'turns' the soil.

But even the turning plow can be handled in a comparatively harmless way. The problem became serious only when farmers forgot that plowing is an 'art', that there are definite and limited conditions under which a soil should be plowed at all. The tractor too, became a tempter to wrong practices with its great power reserve which allows a very deep plowing.

Harm is done to the soil when the live layer on the surface is turned under to a depth where no life is possible and when the subsoil and dead lower layers are brought to the surface where they require years before the microbe and earthworm life can restore a proper humus. The humus turned under will bring about an acid decomposition, forming a solid dead layer, eventually like a kind of peat moss. This all hinders the root development of cultivated plants. The improper plowing of too wet a soil, gradually forms a hardpan impenetrable by air, water or roots, thus increasing the acidity, etc. Also it may turn up big lumps which will lie in the ground for years.

Deep plowing, the great temptation on deep, fertile soils, is to be condemned by all means. If one needs to go deep in order to increase drainage or to break a hardpan one should use a subsoiler, Killefer or something of the sort in addition to surface tillage.

The live layer of the soil is at best from one inch below the surface down to about 3 or 4 inches in a well-aerated soil. The material within this layer is interchangeable without loss or destruction of the life. A disking of this layer is good, also turning within the layer will do no harm. In other words, as long as the plow remains within the live layer and is otherwise handled according to right moisture conditions, (e.g. not too wet and not too dry), avoiding hardpan, lumps or crust, it will do no harm. This proper handling of the plow according to the life conditions and structure of the soil, is the 'art of plowing'. Farmers possessing this art, do no harm to the soil, according to the observation of the writer. Their soil crumbles nicely, does not form a hardpan, in some instances the fields look as if harrowed rather than plowed. The problem is that for many plowmen this is a lost art. Can these men who have lost the art of plowing, appreciate the art of Faulkner? They are not soil conscious and do not understand the way of life in a soil. If Faulkner's book is able to stir these men up, to stimulate their minds and bring the living processes of the soil to their attention, then it has done an important job. In that case we do not need to make a dogma of *Plowman's Folly*, which some persons are afraid may happen, for we will understand how to do the right thing at the proper time and place, even how to handle the plow without doing damage.

°*Bio-Dynamics*, Winter 1944-45

VII

THE BIO-DYNAMIC METHOD CAN PRODUCE QUANTITY TOO

The proponents of the mineral fertilizer concept, of the Nitrogen-Phosphate-Potash, etc., theory point out that the minerals which are removed by a crop must be replaced because the soil is not inexhaustible. They also say there is not enough organic matter available to cover the needs of the soil, even if one were to concede that the organic concept has a proper place in this modern agricultural world. The mineral concept underwent several changes and amendments during the decades after J. V. Liebig originated the basic idea. It has been shown that trace minerals are as important as the major nutrients, N-P-K; in fact 5 lbs. of borax or 1 oz. of molybdenum per acre are as decisive with regard to the yield, for instance, of alfalfa, as the application of 120 lbs. of nitrogen or 500 lbs. of potash per acre. It was discovered that only the proper balance of all factors guarantees the full efficiency of fertilizers. The proper balance, which also includes soil life, organic matter, and soil structure, secures a soil management without losses or the need of excessive application, that is wastes or deficiencies.

Agricultural schools, experimental stations and extension services all over the world emphasize the use of supplementation. It is true that where one-sided crops in monoculture demand from a soil the impossible, supplementation is necessary. Entirely different, however, is the situation when a balanced system is applied. In the following we shall describe the achievements of one group of farms which has used the Bio-Dynamic method in a balanced, self-contained system and which has, as we shall see, fared very well.

These farms have aimed at being self-supporting with regard to fertilization. The principles underlying this aim were simple: ——— (a) To use on the farm only organic matter produced by way of manure and compost from the offal on the farm. To plow under fresh manure is wasteful. Recently it has been discovered that this practice increases acidity and promotes leaching. The application of raw organic wastes is spendthrift too; it leads to considerable losses of organic matter and nitrogen; it spreads diseases, especially fungus diseases, and may even transport toxic elements. The treatment of manure and compost, according to the Bio-Dynamic method, has enabled these farms to preserve all values and to effect the most far-reaching, economical and efficient use of the on-the-spot available organic fertilizers. These farms produced just enough manure and compost to cover 20% of the entire agricultural surface per year. They were not overstocked but had just enough livestock to do this; for instance, 20 head of cattle per 100 to 125 acres of land. _____ (b) Second, the crop rotation was thus organized in such a way that 3 years of root and grain crops were followed by at least 2 years of grass-clover-alfalfa cover. If a grassland management could be achieved, it was that much easier. These farms about which we will report had a crop rotation of potatoes or sugar beets, followed by winter wheat or barley, spring wheat or oats, beans or peas, followed by new grass seeding. Occasionally flax was grown too, especially when the soils offered structural problems, for it was observed that flax improved the soil structure. In the region where these farms are located, flax was economically a good crop. _____ (c) Soil structure, especially on heavy clay soils, was well maintained with proper cultivation, avoiding the production of hardpan or caked surfaces. This requires skillful cultivation and understanding of the properties of the soil, that is, good field management. The art of cultivation, unfortunately, has faded all over the world. _____ (d) The water household is important. These farms were favored by nature with a high ground-water level which could be somewhat regulated. Also, the climate, though cool, was favorable in order to avoid humus losses. Dry seasons, in fact, gave better results than wet seasons, although in recent years the top yields have been in the wet years. It should never be forgotten that water is the limiting factor for top quantity production, and not fertilizers. Without water, no fertilizer will do any good.

The farms we are reporting about belong to the N. V. Cultuur Maatschapij Loverendale and are located: two (TM and TL) on the island of Walcheren, one (PH) in the Oude Prinslandsche polder of the province N. Brabant in Holland. The latter farm is in an area of some 250 years of cultivation and of high soil fertility, with heavy soils while the two other farms are on lighter soils with not too high a fertility level originally. Loverendale was founded by the late Miss Tak van Poortvliet and this writer in 1926 in order to demonstrate the practicability of the B. D. method. Not everything went smoothly and easily; in fact, everything we achieved we learned the hard way. This writer was connected with the company as director for the first ten years. He laid out the policy of the company and did the spade work. Dr. Hans Heinze took over and has guided its development ever since. He performed active, on-the-spot duty until after the war. The director of the company at that time was Mr. M. Steyn, whose understanding of soils and management has brought about the results which we report herewith. But then, Mr. Steyn, from local farm stock, had grown up on the farms of Loverendale and represents the very best peasant farmers' tradition.

The following table contains a survey of yields over a 25 year period:

INCREASE OF YIELD OVER A 25 YEAR PERIOD
AVERAGE YIELD PER ACRE IN BU. PER YEAR

FARM	TYPE OF SOIL	CROP	PERIOD 1935-39	PERIOD 1953-60	TOP YIELD	YEAR
PH	heavy clay	wheat	47.8	57.7	69.0	1960
TL	medium heavy	wheat	45.8	60.8	66.0	1960
TM	light	wheat	45.8	73.7	84.4	1969,'60
PH	heavy clay	oats	89.0	95.9	110.6	1960
TM	light	oats	101.8	166.9	168.0	1955
PH	heavy clay	barley	53.7	75.0	86.0	1959
TM	light	barley	67.9	89.9	921	1955,'56
PH	heavy clay	peas, seed	41.5	46.6	51.7	1955,'56
TL	medium heavy	peas, seed	33.3	47.1	55.7	1960
TM	light	peas, seed	33.3	55.2	62.2	1956

1960 AVERAGE YIELD PER ACRE IN SHORT TONS

PH	heavy clay	sugar beet	15.0	16.3	22.0	1960
TL	medium heavy	sugar beet	15.6	18.1	22.0	1960
PH	heavy class	potatoes	7.7	9.1	11.0	1960

The trend of steady increase of yield over these 25 years of bio-dynamic management is obvious. All figures are confirmed and show up in the financial status of Loverendale. There were fluctuations, of course; some years, because of extremes of weather, the figures were low, but the average still remained high and increased until the top level was reached. The remarkable phenomenon, however, is that no commercial fertilizers were bought during this period; that the war years and a 14 month flooding with salt water because of the war have not destroyed the land. Before the farms were converted to the Bio-Dynamic method, the land was intensively fertilized with commercial fertilizers; in fact, during the first 5 years after the conversion, certain let-downs were unavoidable, but once the soils began to respond to the B. D. treated manures and composts, the trend went upward.

The last two years of the reported period, in many instances, showed the maximum yields, which can be considered high yields in any country of the world. For comparison we report top yield figures from Manitoba, Canada for wheat - 56 bu.; Wyoming - 50 bu.; New York - 60 bu.; for barley, Manitoba - 75 bu.; Washington - 80 bu.; for oats, Manitoba - 80-110 bu.; one, the highest in Washington - 150 bu.; field peas, Manitoba - 46-65 bu.. Most U.S.A. yields are considerably lower (all figures are bu./per acre).

Nitrogen never was a problem on these farms because soil life and nitrogen fixation as well as the conservative crop rotation took care of the nitrogen supply naturally. The farms were typical grain and sugar beet farms, these being the main cash crops. Loverendale had about 300 acres of farm land; before the war, 21 acres of truck gardens and 2-1/2 acres of green houses. In the latter also good yields were obtained. Especially the greenhouses with tomatoes, cauliflower, cucumbers, grapes and chrysanthemums produced very well; no diseases interfered and no insecticides were ever needed.

The development of Loverendale proves that there is definitely another approach to farming besides the usual and common mineral fertilizer concept. What we hear frequently is: 'Well, maybe with the B.D. method you can produce quality but never quantity; to get yield you must use mineral fertilizer'. This is definitely not true, if one really knows how to handle soils with the B.D. method, etc., etc., as Loverendale has done it.

We now have to investigate the conditions under which such high production is possible. Soil analyses 10 years ago and today indicate that the levels of availability (in lbs./acre) remain about the same. The amount of B.D. treated farmyard manure applied over the last 10 year period (1951-1961) corresponds to two applications, each of 10 tons/acre or, in terms of nutrients of about 200 lbs. of nitrogen, 160 lbs. of phosphate and 200 lbs. of potash. Usually the manure was applied to the root crop, either sugar beets or potatoes. The other crops of the rotation - two crops of wheat, barley or oats, one of peas or beans, 3 years of mixed hay - did not receive any extra fertilizer. However, some compost was used from wastes and, of course, all vines, stubble and straw were returned to the soil. In terms of the textbooks of agricultural chemistry, the over-all removal of nutrients is estimated in the neighborhood of 540 lbs. of nitrogen, 262 lbs. of phosphates and 586 lbs. of potash over the 10-year period. This would leave us with a negative balance of 340 lbs. of nitrogen, 102 lbs. of phosphates and 386 lbs. of potash. For this calculation of the removal of nutrients, good averages of nutrient-removal have been used.

The negative balance must come from the soil and its reserves. The problem of nitrogen is simple. With Bio-Dynamic treatment, we can count on the replacement by natural bacterial fixation. It has recently been discussed, in other quarters, that such fixation can take place with other crops too, not only with legumes. Our experience is that 80-120 lbs. of nitrogen per year can be easily fixed as long as one takes care of the soil life and 'feeds' it with organic matter. With regard to the two other nutrients, we can only assume that these come from resources locked up somehow in the soil. Again, according to the mineral balance theory, a decline of yield should show up after a while, definitely not an increase. The facts reported disprove the validity of this concept. This means that the mineral supply concept applies to mineralized soils under conditions as they exist in nutrient solution, but that organic soils behave differently.

The fertilizer theory has been developed because of a concept gained from the behavior of plants in nutrient solutions, in closed vessels or systems. A living soil, however, is not a closed vessel. Other conditions prevail in a living soil, conditions which are evidently not present in a fertilizer nutrient solution (only) system. In BIO-

DYNAMICS No. 57 we have already discussed the problem of the hydroponic-nutrient solution approach. Experimentally it can be shown that even in a nutrient solution-hydroponic culture an entirely different growth of roots can be observed when organic matter in the form of stable humus is added. The plants then develop from two to three times as many roots, which enables them to make an entirely different use of the mineral nutrients. We learn that the utilization factor is entirely different. It is here where our empirical knowledge rests for the time being. Future research should make us familiar with the hidden reserves and the way to make these hidden reserves available. Are these reserves inexhaustible? Or are there conditions in which a soil sustains itself?

In human nutrition there is a similar problem involved in the need for protein, for instance. It has been shown that man does not need 120 gr./day or - as the standard is today 70 gr./day of protein but can get by with 40 gr./day if a proper amino acid balance is maintained. The present standard in the U.S.A. is 70 gr. or, better, 1 gr./kg. body weight, as against 0.36 gr./kg. according to the FAO figure. One thing is sure - had we used the common fertilizer concept, we would have introduced (a) a wasteful practice; (b) fared economically not as well as Loverendale did in recent years. Another truth is very clear: The achievements were possible only in a diversified system of farming. A monoculture will always need supplies from the outside. However, if these supplies are organic and biologically conditioned in order to activate the soil, even the needs and balance of nutrients might look entirely different. There is one other item noteworthy in the history of Loverendale: no insecticides or pesticides were ever needed.

The soil analyses of these farms are also quite interesting, in that they prove the stability of the soils; for instance, field TL No. 12, 1951, rates 180 lbs./acre of potash, and 1961, 220 lbs. of potash; 1951 - 150 lbs./acre of phosphate; 1961 - 300 lbs. of phosphate. The nitrate level has remained steady at 20 lbs./acre with slight fluctuation. The organic matter has increased on many fields from a 2.6% to a 3.4% level or more. Only on fields with more grain crops than usual in the rotation was the organic matter level lowered. The calcium level was high on all fields in 1961 (5600 lbs./acre), while the 1951 average was 4000 lbs./acre and, in a few fields, 2000 lbs. No liming was done except in 1946; after the salt water flood, an application of gypsum was made, which was demanded by the government in order to compact the soils. These had become spongy because of 14 months' exposure to salt water. After the flood, by the way, the soils of TL and TM were completely barren for one year but recovered surprisingly fast. One property of these soils is important: as soon as they rally to the B.D. treatment, structural problems become easier and hardpans can be avoided. This, of course, is conditioned by the proper cultivation methods, in which our dutch farmers are masters. Some of these soils are very heavy clay. These respond more slowly, but the dutch skill does not ruin them by plowing them when the soil is too wet. The much dreaded compaction is out of the question there.

Finally, a word about quality. Loverendale produces a high quality of grain, so high in fact, that the wheat can be milled and baked at the very place, blending winter and spring wheat and using a very old-fashioned stone mill. Loverendale thus produces a whole wheat bread which has found its place in Holland as high quality bread. The baker of Loverendale produces 7,000-10,000 loaves (@ 2 lbs.) per week.

Unfortunately, the war has destroyed the market garden acreage and greenhouses. Only a small greenhouse has been rescued from the debris, which, however, produces year after year an outstanding crop of table grapes - with no disease and no sprays. Also, the newly planted orchard, planted after the salt water flood, gives clean fruit, and with no poisoning sprays used.

What does all this teach us? That agriculture can be successful even if it is different from the orthodox fertilizer theory. We can imagine that this idea is not looked upon with loving eyes by the fertilizer industry. But then, in this age of competition, where the farmer is hard up for survival, he has to look out for himself. The lesson we draw so far is that an economy which respects the biological laws rather than the purely chemical concept enables a farmer to survive. It is a matter of philosophy and orientation of the mind which decides which path to follow. Loverendale has come through the war and its destruction unharmed; in fact, upon recovery it became really productive. Is this not a sign as to where real survival lies? Suppose we have another war -

global war, atomic war: one thing is sure - industry will fail the farmer. Food will always be needed, even if a major part of humanity has been liquidated by then. Should we not allow the pioneers of a survival concept of the soils we trod to have a chance to demonstrate what to do and how it can be done? We are quite convinced that the biological age will replace the chemical age. There is plenty of opportunity for the chemical industry to make a living too, otherwise.

°Bio-Dynamics, No. 59, 1961

VIII

ORGANIC MATTER IN SOIL

What is organic matter in soil?

Roots, leaves, decayed plant parts, decayed microorganisms (of plant and animal origin), leaf mold, green manuring crops, stubble being plowed or disked under, compost, manure, trash.

What is humus?

That part or fraction of organic matter which is digested and transformed by soil life, i.e. mainly by microorganisms and natural oxidation. We can therefore speak of raw organic matter and of humus.

Are there different kinds of humus?

Yes, acid or raw humus and neutral, colloidal humus.

Acid or raw humus (pH below 5.5) is unstable, can be washed out (chemical erosion).

Neutral, colloidal humus is stable unless the soil is flooded or dries out completely over a long period of time.

What does colloidal mean?

Colloidal comes from the Greek word 'Kollos' and means literally glue. Colloid now means that state of half-liquid half solid substance or fluid which exists for instance in the cells or plasma or other body fluid in living beings. It is characterized by its high power of absorption or 'holding power'. In a living cell or body, all organic functions of growth, maintenance, etc., are performed through the colloidal state.

Why is neutral humus preferable to acid humus?

It has a high absorption power, is able to absorb its own weight in water, is teeming with soil life.

Nitrogen-fixing bacteria prefer neutral or near neutral humus as a medium to live from.

It 'holds' the soil, counteracting sheet, gully, and chemical erosion.

It is the chief factor which maintains soil fertility. Without humus there is no stable fertility.

How much organic matter is and should be in a soil?

U.S. average of agricultural (cultivated) soil is at present 1.5%. Fertile, virgin, or other good soils contain 4-6%.

Below 1.5%, a soil is endangered; below 1.0% it is in a critical state.

Good clover fields, alfalfa fields, or pastures may have as much as 3-4%.

Grain crops and truck garden crops consume much humus in intensive cultivation and aeration, unless organic matter is replaced.

Forest soils may have a high organic matter content, but it is mostly acid-raw humus.

Trees in general, also fruit trees and berries, require a lot of humus.

From the organic viewpoint, there should be at least 2-3% organic matter in a soil. Only then is permanent fertility guaranteed.

How can soil be tested for organic matter?

Soil tests are made for organic matter, telling the percentage of it. The test for acidity (pH) then tells whether the organic matter is acid, raw humus, neutral, or rather neutral, etc.

The test for nitrate suggests how much stable nitrogen in available form is present in a given soil.

The bacteria count tells about the amount of bacteria, i.e., how much soil life exists.

From these factors one can conclude how much stable humus and microbiological activities exist in a soil, how fertile the soil may be, how resistant against erosion, etc., and how far one can go with fertilizing, cultivating, irrigating, permanent cropping, or crop rotation without doing harm to the maintenance of fertility.

What is compost?

Compost is organic matter deriving from any kind of organic refuse, garbage, leaves, industrial wastes, straw, leafmold, sawdust, in a certain state of decomposition with or without admixture of earth. The decomposition of all these matters is brought about by microorganisms and by chemical and structural breakdown of the original source materials. For instance, protein breaks down to amino-acids, finally to ammonia, carbon dioxide, etc.

Cellulose breaks down with carbon dioxide being the final product. Starch is hydrolized, etc., etc.

If organic matter ends up in ammonia, carbon dioxide, mineral elements, nitrates, etc., it would mean losses of all carbon and nitrogen molecules and eventually, through washing out, also of the mineral elements.

The art and science of compost making therefore consists of the phases: (1) the breaking down of organic matter to such a degree that other microorganisms can take over so that (2) new stable, humus-like organic products are formed.

The first phase can be compared with the digestive process in animals, which leads to manure; the second phase is a self-contained microbiological process which takes place in touch with soil organism.

Compost therefore can be two entirely different things.

A. *Broken-down Matter.* This is not stable, may lead to further losses of nitrogen and carbon and still needs soil life. It is 'food' for soil microbes, but under unfavorable conditions will tie down this soil life and not become available at once as fertilizer of the soil.

B. *Humus Compost.* This is the most perfected breakdown process by means of which stable humus is produced that does not tie down the micro-life of soil, but is immediately available for soil life and plant growth. (The B.D. Compost Starter produces this type.) It is in the interest of the grower and farmer to get the latter type, because with the first type he cannot get the full benefit of the organic matter at once.

What is manure?

Manure is organic matter digested by animals, but is not yet humus.

It can be used fresh: then the soil life transforms it into humus. What has been said for type A compost holds true here too, especially concerning losses of nitrogen.

It can be applied as compost manure. Then it does not smell like manure (ammonia), but smells and looks more or less like brownish or blackish earth. Only in this case are all substances preserved and nothing lost.

Are losses in raw compost or manure considerable?

Yes, quite; up to 50% nitrogen, 50% carbon, 20% phosphate, 20-80% potassium losses have been observed.

How much 'Loss' is observed in humus compost?

In general, no phosphate or potassium losses; 10% nitrogen, which however is later replaced again through action of nitrogen fixing bacteria; 10-15% carbon. If well done, more nitrogen can be found after a few months on account of the nitrogen fixing bacteria.

What is the fertilizer value of compost and manure?

We use farmyard manure as a basis of comparison. The fertilizer value of manure consists of:

1) Its contents in organic matter, which has a high absorption and holding power and serves as base and food for microorganisms. This represents more than 50% of the actual value of the manure.

2) Its moisture content, which should not be considered just as plain water, but as an important ingredient supporting bacteria life. Also it contains all minerals and some of the organic matter in solution, i.e. in available form (dehydrated and dry manure do not do this).

3) Its content in minerals, such as nitrogen, phosphate, potassium, calcium, magnesium, silica, trace minerals.

4) Its content in hormones, growth-stimulating substances, vitamins, and enzymes. The importance of these factors is evident, for manure stimulates growth. It contains a certain growth-factor which has even a remedial effect on diseased pigs and poultry, healing certain deficiency symptoms and paralysis.

5) Its bulky, aerated consistency, which aerates the soil when plowed or disked under, thus helping to improve the soil structure.

Manure therefore not only increases the soil fertility, but stimulates soil life, and improves soil structure. To judge the manure fertilizer-value only for its contents in nitrogen, phosphate and potassium (NPK) would mean that one forgets about 50% or more of its beneficial actions. The NPK content is important, but does not tell the whole story.

Humus compost in many cases is up to 1½ times as good as farmyard manure, while raw humus or raw compost may be less good than farmyard manure.

Manure: The average moisture content of fresh manure is between 70-80%

The average moisture content of rotten humus or composted manure is 50-70%

The average moisture content of air-dried manure is 30-50%

The average moisture content of dehydrated manure (no life contained there) is 10-20%

The average organic matter content of fresh manure is 17-21%

The average organic matter content of rotted compost manure is 20-40%

The average organic matter content of air-dried manure is 35-40%

The average organic matter content of dehydrated manure is 70-80%

The average mineral content of fresh manure is 3-11%

The average mineral content of rotted manure is 8-15%

The average mineral content of dry manure is 8-15%

(Figures in most literature show a higher mineral percentage, but this is due to admixture of earth or losses of organic matter. There is actually no increase of inorganic matter through drying above the original content.)

The average mineral content in fresh manure is:
    Nitrogen: 0.50%   Phosphate: 0.23%   Potassium: 0.50%
    Calcium: 0.60%   Magnesium: 0.20%   Silica: 1.50%

In air-dried manure:
    Nitrogen: 1.4%   Phosphate: 1.3%   Potassium: 2.3%

In poultry manure (with 56% moisture, 25% organic matter):
    Nitrogen: 1.54%   Phosphate: 1.54%   Potassium: 0.85%
    Calcium: 2.5%   Magnesium: 0.8%   Silica: 3.8%

Compost:
>The average organic matter content of compost is 20-40%.
>The average mineral content of compost is 10-40%
>The average moisture content of compost is 40-60%.
>The average mineral content of compost itemized is:
>Nitrogen: 0.5-0.8%   Phosphate: 0.5-2.0%   Potassium: 0.5-1.0%
>Calcium: 10%   Magnesium: 0.1-1.0%

>*Poor composts analyzed:*
>>Nitrogen: 0.3%   Phosphate: 0.3%   Potassium: 0.3%
>>Calcium: 0.1%   Magnesium: 0.01%

Com-Co Soil Builder:
Nitrogen: 0.1-2.0%   Phosphate: 2-3%   Potassium: 1-2%

Com-Co Fertilizer:°
Nitrogen: 4%   Phosphate: 2%   Potassium: 1%

>(°made with B.D. Compost Starter)

How much manure is commonly applied and considered satisfactory?

Applications of manure are regulated somewhat by the crop rotation, the type or kind of crop, whether the rotation contains rest periods, green manuring, or other use of fertilizer, and considering the soil analysis and fertility. Some crops are manure-exhausting, some are neutral. All legume crops even improve the soil structure, especially the nitrogen content of soil, through bacteria action.

Exhausting crops are: corn, potatoes, sugar beets, mangels (fodder beets), tomatoes, onions, all kinds of cabbage, cauliflower, cucumber, melons, strawberries, all kinds of berry bushes, grape vine, very intensive cropping of small fruit trees. These have high requirements of manure or compost.

General Rule:

>1) If exhausting crops are followed by exhausting crops the soil will gradually run down, no matter how much compost, or fertilizer is given. 8 to 12 tons of manure, 6 to 10 tons of compost would be

needed for each exhausting crop in order to maintain the fertility. Mineral (NPK) fertilizer in such cases can take care of the NPK situation only, but will not solve the soil life (bacteria) problem, the structural problems of the soil, the moisture-holding capacity of the soil and the need for organic matter above and beyond 1 to 1.5%. These factors can be controlled only by manure and compost.

    2) If an exhausting crop is followed by a neutral crop, the gift of manure to and before the exhausting crop will last for said neutral crop too, i.e. for two crops one gift of 8 to 12 tons of manure or 6 to 10 tons of compost is enough.

    3) If the exhausting crop is followed by a neutral crop and this in turn is followed by a legume crop, the original gift of manure or compost will last in its effect for these three crops in succession. This is true for medium to good soils. For poor soils the rotation is more successful if the exhausting crop is followed by a legume crop and the neutral crop comes third.

    4) If the 3-crop rotation is followed by a cover rest, or green manuring crop, the after-effect can be extended into a 4th; on very good soils with a high (2.5-5%) organic matter content even into a 5th year or place in the rotation.

    5) Whenever the soil is used again for an exhausting crop, a new gift of manure or compost is required.

    6) With this manuring and cropping system, mineral fertilizer is only needed in the interval phase as correction of actual deficiency. Later on, mineral fertilizer may be needed only in greater intervals, as correction, if any. This way the fertilizer becomes more effective and will last longer.

We have seen soils which were properly handled in this way and did not need any mineral fertilizer at all for periods of 10, 15, and even 20 years, and still produce top yields.

    7) The need for manure or compost is controlled by the organic matter content of the soil and its microlife.

Soils with 0.5-1.5% organic matter need heavier and more frequent applications.

Soils with 1.5-2.5% organic matter need lighter and less frequent applications.

Soils with 2.5-5.0% organic matter need less at a time at greater intervals.

On the best soils a 5 year rotation may need only one application of 8 to 10 tons, to be followed with 5 to 8 tons at the second intervals.

Soils with 0.5 to 1.5% should be raised in their organic matter level by frequent and heavy gifts first.

A farm with a well established 5-year rotation therefore needs 10 tons of organic matter (humus, compost or manure) and the cost for it is spread over 5 years, i.e. the cost is for 2 tons per year.

8) A specific problem is monocultures, fruit trees, intensive truck gardens. Here, every other crop or year may need compost or manure. If the soils are good, 5 to 6 tons every other crop may be sufficient. The cost therefore is 3 tons per year. Fruit trees, vineyards, and berries do better with a gift every other year and green manuring, winter, or cover crop the other year. The cost again will be spread over two years, ie. 2.5 to 3 tons per year.

9) Truck gardeners, orchardists and citrus growers spend on the average of $80.00 per year in fertilizer (1952). This could be better spent on compost in order to build up the soil. Always remember that one-sided use of fertilizer increases the problem. The soils on this earth which have lasted for centuries and produced healthy crops were all well provided with soil life and organic matter.

IX

HOW MOIST IS MOIST?

In instructions for composting as well as in recommendations for soil management, ploughing and cultivation, the terms, proper moisture, moist but not wet are frequently encountered. The question is asked what is 'proper' moisture and how can I recognize it? Occasionally one gets a letter telling how the writer made compost from straw, dry weeds and/or corn cobs and poured water over the pile so 'that it ran off'. The sample accompanying the letter was analyzed in our laboratory and the writer was told the compost was too dry. In this case the water had run off indeed but had not penetrated, was not absorbed, moistened only the surface but had not entered the core of the particles.

Another question is when is the right time to plough a soil with regard to its moisture content. This problem is especially important for clay, loam and adobe soils since these soils are very sensitive to mechanical treatment. If they are too dry they cannot be handled. If they are too wet they cannot and should not be handled. If they are a little too wet they will compact, cake afterwards and get lumpy if ploughed.

Now we are asked could you not tell us in figures what is moist and what is wet? Is a moisture content of 55% say, the borderline? The only answer to this is that each material behaves differently and that one cannot generalize. In one case 50% may still be rather dry and in another may be soaking wet.

Plants, for instance, contain 80% moisture or more and are still not dripping. The same is true of the human body, even the 'dry' bones still contain a certain amount of moisture.

Our laboratory has set up a little demonstration of the different degrees of moisture. At our conferences and demonstrations we let visitors see and feel these samples of different materials. Anyone who is interested in the problem can arrange a little test for himself. It is with this purpose in mind that we have worked out the data on a few materials in the following table.

MATERIAL	% MOISTURE	APPEARANCE, STRUCTURE, COLOR
Virgin Soil from Missouri with high humus content not touched by man 6% organic matter	8	dusty, grey
	20	slightly cumbly, slightly moist, dark grey
	30	perfect structure and moisture conditions, black, typical humus look, crumbly
	40	begins to lump, black, moist, begins to get sticky
	45	standing, glistening moisture shows up. Wet

The ideal moisture content here is between 30 and 35% and 45% is the borderline of absorption

Adobe Soil from California 3.5% organic matter	10	dusty and hard, caked lumps, whitish grey
	20	loose, crumbling, moist to touch, dark grey to grey-brown color
	30	sticky wet
	40	muddy, water standing, and glossy surface

The borderline for working this soil is near 25%.

Sandy Humus Soil from Arkansas 5.5% organic matter	2	dry, dusty, grey-brown
	10	loose, moist to touch, dark grey-brown
	20	becomes sticky, clay fraction gets lumpy, wet
	30	soaking wet, sticky

The borderline is near 25%.

Clay loam from Chester, N.Y. medium heavy B.D. treated 4.7% organic matter	8	dry, not dusty, crumbling, light brown
	20	loose, crumbly, moist to touch, brown
	30	still crumbly, moist, dark brown
	40	slightly muddy, wet, glossy surface

Borderline 35%.

Organic materials such as peat, manure, and compost are quite different in their moisture behavior.

Peat 73% organic matter	10	dusty, dry, grey
	20	dusty, dry, grey
	40	slightly moist to touch
	50	slightly moist, changing color from grey to brown
	60	moist, granular like coffee grounds, changing color to black-brown
	70	crumbling, spongy-moist, black
	80	wet, water standing on top

Borderline of absorption 75%

Fresh cow manure no litter	40	slightly moist, yellow brown, spongy to touch
	50	moist, spongy, lumpy, dark brown
	60	moist, clumpy, dark brown on side exposed to air, green-brown inside
	70	moist to wet, green-brown
	80	wet, sticky green, lumps
	90	real wet, moisture visible, water oozes out

Borderline is 75%

Cow Manure Starter treated well fermented	20	dry, yellow-brown to brown, particles fall apart easily
	40	moist to touch, dark brown, particles hold together
	60	moist to touch, dark brown, particles begin to stick together
	70	moist, but still absorbs moisture, black brown
	80	moist to wet, moisture shows and can be squeezed out

Borderline of absorption is 85%.

Compost from garbage half rotted, 50% vegetable waste, 35% soil, 15% manure	30	moist to touch, lumps, dark brown moist, but still absorbing water, lumps, black-brown
	50	wet, glossy, water shows and can be squeezed out, green brown

Borderline of absorption is 45%.

Well-rotted compost same mixture	10	dry, dusty, powdery, grey-brown
	20	still dry, powdery
	30	moist to touch, crumbly, dark brown
	40	moist, crumbly, black brown
	50	wet, moisture shows, water can be squeezed out

Borderline of absorption 45%

As can be seen from these examples, an increase of moisture darkens the color but the borderline of absorption is quite different, so that the color alone does not give the exact amount of moisture unless one is familiar with the specific properties of the material. In general it can be said that the amount of organic matter increases the absorption and raises the saturation level in soils and composts, provided the organic matter is digested and transformed into humus. In fact, the degree of humus formation can be judged from the absorption factor. Peat behaves entirely differently from all other materials.

In a compost mixture the absorption will increase with fermentation. Very well-rotted composts with a high percentage of organic matter, containing less that 25% soil (not a 'dead' soil) can absorb as much as their own weight in water, i.e. a ton of such compost will absorb as much as 250 gallons of water as we have seen in our experiments. Soils low in organic matter usually absorb very little moisture and dry out quickly, sandy soils especially, while clay and loam soils hold somewhat more moisture.

Every farmer and gardener can make similar demonstrations of his own soils and compost on his own place without the help of a costly laboratory. While these home and field tests will not give exact quantitative figures, they will allow a certain qualitative evaluation. In this case, where no moisture determining apparatus is on hand one can proceed as follows:

Take a measured sample of your soil or compost, 4 ounces or more (for easier calculation 100 grams equal 3.6 ounces). You can weigh it on a postal scale. Take the sample as is, but describe its appearance, structure, color. Spread it on a dish or plate in a thin layer and let it air dry (a damp day with high air moisture is not suitable) or place it in your kitchen oven at 220 degrees F. for one or two hours. Weigh it again and see how much weight equals moisture it has lost. Put the dry sample in a small cup. Use a graduated measuring cup or cylinder and add water until the sample is real moist, gets a glossy surface and does not absorb any more. Add water slowly. Weigh the wet sample and see how much the increase is in ounces, grams or cubic centimeters of water. Or read on the graduated cup or measuring cylinder how much water the sample has absorbed. Compare different samples of different soils and composts. The higher the absorption the better the organic status of your sample. After you have made a dozen or more of these simple tests you will develop quite a good amount of knowledge with which to answer the question: How wet is wet?

Bio-Dynamics. Vol. XI, No. 4, 1953.

MULCHING - RIGHT OR WRONG?

Mulching is an almost ideal practice in farming, and even more so in gardening, in order to protect a soil against excessive sunrays and drying out, against losses of humus because of intensive oxidation (especially in southern, warm climates), and against losses of moisture. The cover of a forest soil, with leaves or needles which decompose slowly underneath the shade of the trees, is a natural mulch process. This natural cover of forest soils teaches us all the good aspects of mulching, especially moisture conservation, but also all the dangers of a permanent mulch cover; the separation from air and therefore increase of acidity, water-logging, one-sided microflora with preference given to fungi, stratification underneath and eventually podzolization. The prevalence of humic acids is definitely a double-edged sword. One-sided mulch as, for instance, in a monoculture of conifer, from pine needles, is rather hostile to many other plants. New seedlings will settle and flourish only if this mulch is occasionally scratched, broken up and interrupted and the soil underneath aerated.

From all these observations, we already learn that mulch is a remedy at times but can become damaging at other times. Those who want to mulch successfully need to make themselves familiar with a few basic rules and need, especially, to understand the purpose of mulching and when and for how long it has to be applied.

These are the basic rules: if a soil is well drained, the mulch will protect the soil against evaporation of water, that is, water losses. Underneath in the cool, moist atmosphere, humus will be preserved and more humus will be produced. This mulch should not be so thick as not to let air circulate and eventually rain penetrates. This mulch is ideal in a dry and hot season. In wet and cool seasons it can do great damage by collecting and retaining water. Such a mulch in an orchard in a wet season can do a lot of damage, as was reported to us during the summer of 1958. Many a mulch fanatic (there are such people) has blamed everything else as the cause of trouble in his orchard, while, had he removed the rotting mulch cover, he would have solved most of his problems. When told to do so, he answered: 'Well didn't you say mulching is important?' 'Yes, my friend', I reply, 'but under conditions.....'

We differentiate between two kinds of mulches - the nourishing mulch and the protective mulch. The nourishing mulch is a soil cover to last for a certain period of time and then to decompose gradually. This mulch, in order to be able to decompose completely at the end of its purpose or function (latest at the end of the season), should be a compost in a certain state of breakdown, most likely a half or three-quarters broken down compost. This means that the compost is not 'raw', showing a structure, odor and appearance of any of its raw sources - manure, straw, leaves, garbage or whatever. The first smelly breakdown phase should be over. On the other hand, this compost should not be so advanced as to be earthy because then it can not serve as mulch any more. Should this compost contain a lot of fibre, as in leaves or straw, so much the better for its task as a mulch. This compost can be enriched; that is, it can contain a sizable amount of nitrogen, phosphate, potash, trace minerals or lime wherever required. Half rotten leaf mold makes an excellent material. Tall weeds with strawy stems make excellent material, provided that they have been cut before they have a chance to go to seed. Mustard, ragweed and stinging nettle make excellent nourishing mulches. At the end of their task or of the season, the leftovers can be worked into the soil surface by any method of cultivation (plowing, harrowing, digging, hoeing, etc.) In orchard management with no cultivation there will be usually a good grass sod development. The grasses can be mown and left as hay mulch, especially in cases where it is planned to change from no cultivation to open soil cultivation. It should be kept in mind, however, that any cut grass or hay cover may choke out the grasses underneath. This applies even to grass cuttings left on a lawn. If the aim is to maintain the grass cover, it will be necessary to rake the hay together and to remove it before the grasses underneath are choked, that is, get yellow and begin to die.

The protective mulch: this kind of mulch has the sole purpose of protecting the soil against moisture losses and hindering evaporation of water. These mulches will be especially beneficial during dry periods. It is evident that these will perform best if applied while there is still moisture in the ground. Anticipating the weather is frequently guesswork. One will therefore make certain errors, namely, apply the mulch and then have a long period of moist and rainy weather. The mulch may rot, or an acid, waterlogged condition may develop underneath. In such a case, the mulch should be loosened up, should not remain packed and eventually might even need to be removed.

Protective mulches give little in the way of nutrients to the soil. Such mulches can be hay, straw, weed straw, leaves, bagasse, even paper, polyethylene or other plastic materials which can be rolled on and off. Stone mulch would also be classified in this category.

All mulches properly applied will foster the development of a very fine humus and crumbly soil structure underneath. On sandy soils, the water retention will be especially beneficial. On very heavy clay soils which offer considerable structural problems and are hostile to cultivation, a mulch may be the only way to render them crumbly. In such cases, the clay should be reasonably dry and the mulch should not be applied too early, that is, not on wet clay.

The dangers of mulching: The greatest danger consists in the fact that mulch hinders the access of air and that a waterlogged condition may develop. Many mulched orchards during the wet spring and summer of 1958 encountered more trouble and harm than good, and the development of fungus disease became excessive. If a mulch gets moldy, this means it was too wet, then the soil will become acid under it and no nitrogen assimilation is possible. Such mulches should be removed before they become harmful. This means that mulching is not a practice to be followed stubbornly under all circumstances but should be done with good common sense and an understanding of its function. The effect of mulching depends on the observation of the conditions of soil and weather - wet or dry.

Another danger exists when mulches are left over the winter. If the soil is well covered with any fibrous, loose material (hay, weeds, leaves), these might become the ideal hiding and hibernating quarters for insect pests. Mice and rats just love to tunnel under the protection of this kind of mulch; rodent damages to the lowest part of the trunk of a tree go unnoticed. Soggy mulch also is a good breeding place for fungi.

Under climatic conditions where frost plays a role, there is nothing more natural and better than frost action on a soil to produce a crumbly structure, provided the frost can freely penetrate and the soil is well drained. A waterlogged, soaky mulch which freezes and thaws is of no help and only acts as an insulator; does not even keep the soil warm and does not protect against frost damage to sensitive plants. The freezing out of plants during the winter is increased with waterlogged conditions of the soil or mulch.

Even snow acts as mulch. Underneath the snow, the soil warms up to about the depth that the snow is thick. Two feet of snow would penetrate with its warming effect about two feet deep.

We hear many an orchardist complain about curculio damage. This beetle is a very nasty one, difficult to combat, especially if one does not want to apply poisonous sprays. In letting mulches lay over winter, one creates the best condition for its hibernating. If it gets the upper hand, I see no other way but clean cultivation over the winter and jarring right after blossoming time, at the first mating season. One spreads a paper underneath the tree and shakes the branches thoroughly early in the morning at sunrise when the beetles are still cold. In a small orchard where every tree is a beloved pet and treasure, this is not too much work. Heinz Grotzke recently showed me how he could collect 200 curculio in a few minutes. This, done for several days will decimate the pest. If the trees, leaves and the setting fruit have been covered with the after-blossoming tree paste spray, one has a good chance to control the curculio. This is then a question of proper 'sanitation' of the orchard's ground floor.

Finally, no mulch should come close to the trunk of a tree. There should always be an open collar of one to two feet around the trunk. This 'collar' can be level or slant away from the trunk, never towards it.

For all the reasons described, we feel that alternating of cover crops, (sods), with open cultivation and mulching during the critical season (dry season) might offer the best sanitation.

Mulching against weeds has to be done with a protective mulch, thick enough or impenetrable enough so that weeds are choked out. Chopped straw or bagasse are excellent for this purpose; even paper or plastics. These should lie long enough to hinder the weeds from growing. Straw or layers of leaves as mulch for this purpose need to be at least two inches thick.

Grapevines, cane and bush berries and strawberries respond very well to mulching; in fact, all row crops where the distance between rows is reasonably wide. Recently we have experimented with the application of the tree paste sprayed directly on the ground, between rows, and even covering the plants while they are young. This has offered an excellent protection; rain penetrates through this cover but evaporation is hindered and good humus forms underneath. This spray can be very thin and will last many weeks.

Another method of mulching is the pouring of a paper pulp mulch on perfectly level land. This, again, offers a perfect cover. This method is applicable to small gardens but rather costly for large areas, for a tremendous amount of paper is needed to get a pulp cover of, say, 1/2 to 1 inch. One builds a frame around the bed and pours the liquid mulch into it.

In southern dry and hot climates, mulching during the summer is most essential, especially if no irrigation water is available. Care should be taken that such mulches do not become a fire hazard, as is the case with a thick layer of dry straw, hay or paper.

To sum up: mulching - yes, by all means, but with the use of good common sense.

Bio-Dynamics. No. 51, 1959

XI

OBSERVATIONS OF BIO-DYNAMIC AND MINERAL TREATMENT OF SOIL

Work with an experimental plot was begun in order to compare long term soil changes after treatment with bio-dynamic manure and compost, with soil changes effected by mineral fertilizing. We were especially interested to find out whether the different soil treatments influence crops in regard to health of stand and vitamin content. The experiment is now in its fifth year.

The BD (bio-dynamically treated) section was fertilized in April with a mixture of well-fermented compost (3 parts) and manure (1 part). The amount was carefully calculated in order to apply approximately equal gifts of nitrogen and phosphate in both cases. The M (Minerally fertilized) section was treated on the same date with a 6-10-4 fertilizer at the rate of 500 lbs/acre. The corn and tomatoes in both sections later received a side dressing of the same amounts. The preparation of the seed beds and the sowing were again somewhat delayed by continued rains. However, it was possible to start things earlier this year than in 1950.

A winter crop of rye in both sections was spaded under in mid April, the earliest possible date. One interesting observation could be made in April still. Both the beets and carrots on the M side had to be reseeded because they had germinated poorly and were also washed away by the rains. The seeds in the BD section were not washed away. The corn, tomatoes and beans were hilled and the common practices of weeding, thinning, and hand cultivation were carried out. All seeding of BD and M crops was done at the same time in all cases. The cultivation of both sections was also done on the same date. On June 10th it was observed that the BD beets and carrots were far ahead of the M beets and carrots (even when taking into account the reseeding that had been necessary). The carrot plants in the BD section were 4 to 5 inches high against the 2-1/2 to 3 inch height of the M carrots.

The differences in soil structure were very evident at all times. The M side was washed badly, the BD side not at all. When the rains stopped the M soil crusted, the BD did not. The color of the M soil was light brown, of the BD soil dark, black-brown. A difference in weed growth was also observed. The M section had many, small, stunted weeds; the BD section had a few larger weeds. The following weeds grew in the M plot: carpetweed (Mollugo verticillata), milkweed (Asclepias), smooth pigweed (Chenopodium album), and some Galinsoga (Carduaceae). The only weeds found in the BD section were ragweed (ambrosia elatior), Galinsoga and chickweed (Alsinaceae).

The bio-dynamic spray preparation No. 500 was applied on April 17th; the other spray No.501 was applied on June 19th. The yields of corn, beans, and tomatoes were good in both cases. The corn in the BD plot produced slightly more. Both beets and carrots suffered from the rains. The peppers presented the most striking contrast. The BD plants grew, flourished well, and produced fine, fat peppers. The M plants were dwarfed, stunted and produced dwarf peppers. There was almost no trouble with insect pests. No dusting against the Mexican bean beetle was necessary. Nor were any insecticides of any type used in either section.

The interesting part of this experiment is the observation of the changes in pH and organic matter content of the soils and in the vitamin A content of the crops. In spite of annual fluctuations due to the weather, the pattern remains and is clearly visible. This is true as regards to the differences between the M and BD groups in general, as well as in the case of the different crops. The following tables tell the story and but little comment is needed. All soil samples were taken and analyzed after harvest, that is, the findings given represent autumn values.

TABLE 1

Changes of Acidity, Measured in pH

Crop Area	1948	1949	1950	1951
Blank Soil (untreated)	5.0		4.5	5.0
Bean M	6.3	5.0	5.5	4.8
Bean BD	6.4	7.0	7.0	7.0
Corn M	5.8	5.5	5.0	4.5
Corn BD	7.2	7.0	7.5	7.9
Tomatoes M	5.9	5.5	5.0	5.0
Tomatoes BD	6.8	6.0	7.0	7.5
Carrots M	5.9	5.0	5.5	4.5
Carrots BD	6.8	7.0	7.5	6.7
Pepper M	6.1	5.0	5.5	4.8
Pepper BD	6.7	6.5	7.5	7.1

The pH of the entire plot before the beginning of the experiment in 1948 was 5.8.

The organic matter values given in the next table were determined by the La Motte method (oxidation with potassium dichromate in sulphuric acid and titration with ferrous ammonium sulphate).

TABLE 2.

Changes in Organic Matter Content, in Per Cent

Crop Area	1948	1949	1950	1951
Blank soil (untreated)	3.07	2.00	1.05	2.50
Beans M	3.80	2.60	2.60	2.50
Beans BD	5.80	3.60	3.70	2.80
Corn M	3.60	2.50	2.50	3.30
Corn BD	3.55	3.20	3.30	5.00
Tomatoes M	2.60	3.20	2.90	2.60
Tomatoes BD	5.80	4.20	4.00	3.00
Carrots M	4.10	2.50	3.30	2.10
Carrots BD	4.95	3.20	4.30	2.29
Peppers M	3.57	3.40	3.10	2.20
Peppers BD	5.75	4.00	4.10	3.10

In the M as well as in the BD plots the organic matter content was greater than that of the blank soil due to the beneficial influence of the green manure crop of rye during the winter of 1950-51. The blank soil had only 1.8% organic matter in April before the rye was spaded under. The M soil had 2.1%, the BD soil 2.4%, at that time. The increase of organic matter during the summer in the case of the blank soil is also due to its having lain fallow. In the carrot area in the M section no increase was observed in 1951 and in the pepper area only a very slight one. The higher values of 1948 reflect the fact that the plot had been in grasses and clover prior to 1948 for several years, and a high amount of organic matter had accumulated. Then too, the weather conditions of 1948 were much more favorable than those of 1950 and 1951.

TABLE 3.

Changes in Nitrate Content in Pounds per Acre
(The La Motte test was used to determine these values)

Crop Area	1948	1949	1950	1951
Blank Soil (untreated)	5			16
Beans M	100	120	16	6
Beans BD	80	140	32	32
Corn M	120	70	14	16
Corn BD	240	40	28	60
Tomatoes M	105	8	10	2
Tomatoes BD	120	80	4	2
Carrots M	120	80	4	2
Carrots BD	90	24	12	5
Peppers M	80	100	14	2
Peppers BD	420	24	4	20

Nitrate nitrogen is only a fraction of the nitrogen content in soils. There are other forms of nitrogen which are often present; ammonia, nitrite and organic nitrogen. Soil tests of BD treated soils in general show the absence of nitrites and a very low or trace content of ammonia.

It was therefore of interest to determine the total nitrogen content of these soils, since we find no nitrites and ammonia only in traces up to 5 parts per million. The total nitrogen (minus the nitrate content) is represented by organic nitrogen mainly contained in humus and in the microflora as well as the microfauna of the soil. We regret that these tests were not made prior to 1951; however, they will be continued from now on. Here are the figures for the total nitrogen content in pounds per acre. The original analyses were expressed in per cent by weight of the air dry sample. The Kjeldahl method was used to determine the total nitrogen content.

TABLE 3a

Total Nitrogen in pounds per acre in 1951

Blank Soil in April	4000	Tomatoes M at harvest time	1120
Blank Soil in October	1620	Tomatoes BD " "	3400
Beans M at harvest time	6600°	Carrots M " "	2600
Beans BD " "	5600°	Carrots BD " "	4000
Corn M " "	7200	Pepper M " "	1000
Corn BD " "	9400	Pepper BD " "	4200

°*Nitrogen fixed in root nodules not included.*

It can be gathered from these figures that the nitrate content of a soil is quite different from the total nitrogen content. In none of the cases cited would we consider that there was a total nitrogen deficiency. The blank soil, however, shows great nitrogen losses due to exposure. The BD soils have been able to build up nitrogen reserves of varying degrees. Contrary to general belief, the corn acted as the best preserver of nitrogen, while the beans showed quite contradictory results. It must be repeated here, however, that the nitrogen which was stored in the nodules of the bean roots was not determined. As in previous years it was observed that the BD beans produced a much larger number of legume nodules than the M beans. In reply to the critics, it may be stated that it would be quite a task to collect all the nodules produced in a field test for analysis. It should be done, though, in order to complete the picture. (1)

Bacterial action in soil will release nitrogen for absorption by plant roots, if the **right** kind of bacteria are present. The presence of a well balanced soil bacteria life in humus is, therefore, of the greatest importance in a) the release of the necessary amount of nitrogen and b) the conservation of nitrogen reserves. It has been our observation that bio-dynamically treated composts and proper bacteria cultures, added to soils, will influence the slow and steady release of nitrogen. This matter will be dealt with in another paper in the future.

The available nitrogen in form of nitrates dropped quite sharply from the 1948 level. At that time, the soil was fallow after a long period of grass cover. Some of the organic reserves had been made available through aeration and cultivation. These were used up or washed out and represent losses. We see, however, that the losses are quite different and the availability changes much more unfavorably in the cases of the mineral corn and tomatoes. The BD corn and beans held much better.

In our experience with BD fields on farms, we have not found such drops in the nitrate contents over the years. In fact, the nitrate content has been held at a very high level, much above average findings. (See *Bio-Dynamics*, Volume IX, No.1: 'Soil Improvements with the Bio-Dynamic Method'). The explanation can be found in the fact that the manure-compost mixture which was used in the experimental garden had a total nitrogen content of only 0.65%, while in the straight manures (unmixed with compost) as we used them on the fields, the nitrogen content was much higher; and besides, a higher tonnage per acre was applied. Then, too, in field tests on farms at least one to two years of legumes are included in our crop rotation, while no such rest periods were allowed in the experimental garden.

One weak point of many trials with the so-called organic method in the past (by friends as well as enemies) has been that the nitrogen content of manures and composts used was not determined. Now there are great differences between compost and compost and manure and manure. We have analyzed hundreds of composts, handmade, farm made, and commercial products. There are as many varieties of compost as there are fertilizer brands and formulas. Composts range from 0.4 to 1.0% total nitrogen. Unless someone knows exactly how much nitrogen his compost contains, he may produce failures with compost applications too. Many comparative tests made by opponents report that there was no difference in yield between fertilizer soils and organic soils. Did they take into account the fact that a compost application has to be as scientifically applied to the specific type of soil and crop as a fertilizer application? There is no doubt, for instance, that a compost without nitrogen fixing bacteria will lose out in nitrogen just as much as a fertilizer will. The organic farmer should learn to work out from a solid scientific foundation in his practices and then the opponents could learn something from his experience. Up to the present moment, we observe all too frequently that the 'mineral' opponent carries out the mineral part of his tests very correctly, but evidences considerable ignorance about the organic part. We do not blame him. How can he know as long as the foundation for the organic science is still in its infancy and as long as even otherwise recognized soil scientists and soil bacteriologists reveal an absence of knowledge and experience in these fields?

One thing is sure: in our laboratory many years have been spent in the study of composts, humus, soil and compost bacteriology, and when the experience has been applied under farm conditions, the farmer has been able to obtain top yields without failures or decline in his soils. The fact remains that on BD farms no nitrogen problem exists. The fact should also be considered that organic nitrogen is two to five times more effective than the same amount of inorganic nitrogen, but only - if and when the proper microflora is present in compost and soil; that is, if care has been taken to maintain a high level of nitrogen fixing bacteria and to include legumes in the crop rotation. The writer used to preach the gospel of legumes in the crop rotation twenty years ago and sees with satisfaction that it has now become common knowledge.

How complex the matter is in regard to nitrogen is also recognized by others. We quote from 'Fertilizer Recommendations Based on Soil Tests', by O J. Coleman of the Soil Department of the University of Missouri which appeared in *Better Crops With Plant Food,* Volume XXXV, Number 9, November 1951:

'Unlike mineral nutrients, nitrates are mobile and are carried down into the soil by percolating water. Because of this, nitrates cannot be stored as such in the soil and do not lend themselves so well to measure by soil tests. The amount of this nutrient present in the soil in turn is affected by soil tilth, drainage aeration, the amount and kind of organic matter, soil texture, etc. Nitrogen, however, is held in the soil in a rather stable form as organic matter or humus. A much better estimate of the nitrogen that will be released during the growing season can, therefore, be made by testing the soil for its percent of stable organic matter. As this organic matter decomposes, it produces the mobile nitrates which diffuse readily through the soil with the percolating water to the roots, and as a result they are approximately 100% available.'

Coleman also mentions that, under favorable conditions in the Missouri area, 5% of the total organic matter consists of nitrogen. With a 2% organic matter content, this would be 2000 lbs. per acre, a figure which we have also found in our total nitrogen analysis of soils.

The drop of nitrates in our experimental lot, from 1948 to 1950, as reported above, can be a serious and important warning of the necessity of maintaining the nitrogen balance in a soil. Here we are in complete agreement with the orthodox point of view although we try to achieve this end by organic means. The importance of this is underlined by the fact that there is not enough nitrogen commercially available to cover the nitrogen needs of our land. Then too, the nitrogen fixing bacteria can increase the nitrogen content of soils to a much higher level, and do this much more cheaply, than it could be done by using commercial fertilizers in the amounts which the farmer could afford to buy. A third factor lies in the characteristic behavior of the nitrogen fixing bacteria. They will fix nitrogen in a medium low in it, but they will consume or waste nitrogen in a medium which is high in it. The farmer thus has the choice of either buying nitrogen as long as his money lasts, or of introducing natural organic practices which will produce nitrogen without much additional cost.

TABLE 4.

Changes in Available Phosphate in Pounds per Acre

Crop Area	1948	1949	1950	1951
Blank Soil (untreated)	75		90	65
Beans M	400	190	180	125
Beans BD	150	150	150	125
Corn M	237	200	150	150
Corn BD	103	180	130	200
Tomatoes M	158	200	100	150
Tomatoes BD	150	190	100	140
Carrots M	400	150	100	150
Carrots BD	190	180	125	140
Peppers M		200	100	156
Peppers BD		100	150	160

Changes in available phosphate are mainly due to weather conditions; both dry and wet seasons having a considerable influence. There are also seasonal fluctuations in availability. It is for this reason that we only test and compare samples which are taken at the same time of the year. The fallow of 1948 before the start of the experiment affected the soils so that at that time an extremely high available phosphate was observed. When the plots were under continuous cultivation and cropping, part of the available phosphate was locked up. The decline in later years is not necessarily a result of cropping, but may be due to other soil changes. The answer to this question could only be given on the basis of parallel testing of total phosphates in the soil. Only then could we know whether there had been an absolute decline in phosphates or simply an exchange between total and available phosphates. As a consequence, we began to test for total phosphate content in 1951 and shall continue this in the future. The determination of total phosphate was made with the official method as outlined for fertilizers by the AOAC. Here are the results in pounds per acre, assuming that the weight of the top 6 inches of soil is 2,000,000 pounds. The analytical data were originally obtained in percent of the weight of the air dry sample.

TABLE 4a

Total Phosphates in Pounds Per Acre

Blank Soil	2000 Tomato Soil M	2000
Bean Soil M	3000 Tomato Soil BD	4000
Bean Soil BD	6000 Carrot Soil M	4000
Corn Soil M	2000 Carrot Soil BD	4000
Corn Soil BD	4000 Pepper Soil M	2000
	Pepper Soil BD	3000

From these figures it can be seen that the soils were not depleted in phosphates and that the differences are only in availability. None of the plants in any group showed phosphate deficiency symptoms as can be detected by leaf analysis. It can also be seen that some crops, such as beans, corn, tomatoes and peppers, were able to store up considerable reserves in the soil. The fertilizer and compost had provided a sufficient amount of phosphate for the growth of the plant.

These observations point toward a new avenue of research, namely the determination of availability in the case of various different treatments of soil and the determination of phosphate storage or waste under different systems. It is known and has been made public by competent sources (the Beltsville Maryland Experiment Station and others), that only a small fraction of phosphate fertilizer (10% on the average, with a minimum of 2% and a maximum of 20%) will be taken up by growing plants, while the rest is locked up in the soil or is lost. A revision of the usual conception of phosphate fertilizing is necessary in view of these facts, and organic, biological, and other factors should be discussed and investigated so as to arrive at a better understanding. As our soil tests of BD farms over a period of years have shown, there is a rather constant level of available phosphate no matter how much phosphate in the form of manures or fertilizers has been applied. In other words, in no case has the soil been depleted in phosphate by organic management. To cast more light on this rather complex matter, we quote from an article by S. R. Aldrich of Cornell University, which appeared in the *ORANGE COUNTY FARM BUREAU NEWS* of March 1951:

'Some of the best dairymen who use 20% superphosphate in the stable have been supplying from 50 to 80 pounds of phosphate per acre each year for the past ten to twenty years. The recommended amount is thirty pounds per acre per year. Seldom more than 20 is removed in a crop in one year.

'Rapid soil tests indicate that nearly 10% of the fields on dairy and general farms contain enough phosphorus, so the farmer cannot expect to see a response on the current crop.

'There is no detrimental effect from applying excess phosphorus, but it is not economical, at least not in terms of the short-term crop response. On most farms the money could better be invested in more lime, potash, nitrogen, or in some other farm need.'

This season, as in previous ones, a marked difference in the vitamin A (carotene) content was observed. The carotene was determined after chromatographic extraction with a spectrophotometer against standard solutions.

TABLE 5

Carotene Content in Gammas Per Gram of Sample
(1 gamma equals one thousandth of one milligram)

Crop and Plant Part Tested	1940	1950	1951
Corn Leaves M	54.0	50.0	58.5
Corn Leaves BD	91.2	90.0	94.5
Corn Seed M			54.0
Corn Seed BD			103.5
Tomato Leaves M	35.1		58.5
Tomato Leaves BD	47.2		72.0
Tomato Fruit M		51.5	81.0
Tomato Fruit BD		81.9	101.3
Carrot Root M		18.0	63.0
Carrot Root BD		20.0	81.0
Pepper Leaves M		90.0	54.0
Pepper Leaves BD		110.0	83.3
Bean Leaves M	13.5	12.0	9.0
Bean Leaves BD	56.2	45.0	45.0

In order to arrive at findings which may truly be compared, one selects leaves of equal size, age, and degree of maturity and seed or fruit of equal size and maturity. Annual fluctuations are clearly discernible. The wet summer of 1950 had a depressing effect on some vegetables. Throughout, however, the BD vegetables have much higher carotene values. These results are in complete contradiction to other publications where it is stated that no differences in vitamin content were observed in the course of experiments testing the effects of inorganic and organic treatments.

We have encountered such statements quite often in recent years. They are used in support of other statements to the effect that there is no advantage in the use of composts over fertilizers. The question is, of course, what kind of organic treatment and compost was used. To simply use 'compost' and not pay attention to the NPK in said compost, or to neglect the wide variations in quality and fertilizer values of composts resulting from different methods of bacterial breakdown, can lead to such statements that 'no differences' were observed. With all respect for the sincerity of such investigators, it must be stated that they probably don't know all about compost as yet, nor are they aware of the science and art of directed compost fermentation. There is a great difference between 'chance' decomposition and a directed fermentation. Certain soil bacteria develop vitamins such as B12 or other growth stimulating factors. Others, even though they may break down organic matter, do not develop such growth factors. This, in turn, brings about differences in the root system and growth of vegetable crops and hence the differences in vitamin and protein content.

Further tests to substantiate our statement are in preparation and will be published in due time. This much can be said at present however: The vitamin content of crops grown with different organic fertilizers, composts and manures, whether treated or untreated, can fluctuate as much as 200%.

To write as Professor S. A. Waksman did recently in *BETTER CROPS WITH PLANT FOOD* (Volume XXV, Number 9, November 1951) under the title, 'Concerning Bio-Dynamic Farming and Organic Farming', is to write inaccurately to say the least. He says there, on page 42:

'The biodynamic groups are actually offering for sale a mysterious biodynamic substance to be used as an inoculum for composts. The fact that equally good composts of stable manure can easily be prepared without such complementary substances is overlooked thereby. All that is required is addition of certain mineral fertilizers, in order to obtain excellent composts from plant residues, such as straw, cornstalks, leaves, and other vegetable refuse.'

Our experience has been quite the opposite of this. Waksman, by the way, contradicts himself on page 24:

'The process of decomposition of plant and animal residues - leading to humus formation is highly important in making soil fertile, since the continuous stream of carbon dioxide, ammonia and nitrate, and phosphate, resulting from such decomposition is highly important for continuous plant growth.'

I have the greatest respect and admiration for Professor Waksman's work. When it is a matter of using bacterial cultures to influence and direct the fermentation of compost, he ought to know better. He has some of the important soil bacteria right there in his laboratory. If proper tests were conducted, he might learn differently, if he so wishes. How could we break down city garbage with our bacterial treatment at the rate of 100 tons per day within a few weeks, without bad odors, without losses of ammonia, and create an entirely new fermentation industry, were Professor Waksman's statement true? Even someone who knows nothing as yet about directed compost fermentation must needs admit that certain well-known and well-described bacteria and yeasts play an important role in other fields of fermentation, in cheese, in milk (yogurt and Acidophilus), in wine, and in bread. Modern techniques in these fields do not permit accidental inoculation to take place but introduce quite specific cultures in order to achieve their specific ends. American wines of a Bordeaux, or Rhine, or Burgundy, or Riesling type are only possible because certain specific cultures have been used. Bread made from dough inoculated quite by chance with wild yeast would be quite unpalatable. It is the typical baker's yeast which makes the dough rise and results in edible bread. To deny, therefore, to the composting process, that such specific microorganisms could be found and used which direct its proper fermentation is a retrogressive and not a progressive concept - inasmuch as experience has already shown that it works, wherever and whenever adequate knowledge and skills are applied.

It is regrettable that some of our greatest authorities want to stem the tide instead of getting busy and discovering something perhaps much better than that which we have already: namely, methods to produce higher vitamin contents, to preserve and increase nitrogen resources, to increase organic matter - in short, to raise the level of quantitative and qualitative agricultural production. We should never forget that to stand still is to make the first backward step, while progress is made by searching for ever new processes and new methods. That which was good yesterday may already be inadequate tomorrow.

It is my pleasure to thank all those who contributed to these experiments: my helpers in the garden and laboratory (Mrs. Alice Heckel, Mrs. Erica Sabarth, Mrs. Mathilde Vibber), and the Bio-Dynamic Farming and Gardening Association and other friends who contributed the necessary funds.

----

(1) See *BIO-DYNAMICS*. Volume IX No. 3: Compost and Fertilizer on Peas.

XII

THE TREATMENT OF SOILS WITH REGARD TO HUMUS AND STRUCTURE

Soil is a living organism. It is also a dynamic system of various factors which support or hinder one another, maintain or balance or lead to a decline unless the disturbing factors or deficiencies are counterbalanced again. This dynamic system is exposed to a steady flow of events, is constantly changing. There are for instance, seasonal changes influenced by temperature and moisture. These follow a certain cycle which is not only reflected in the amount of microlife in the soils, but also in the availability of minerals, and especially in the structure of soils. Much has been said in recent years about the chemistry of soils, about humus. Every farmer is told by his agricultural agencies which fertilizer to apply. How to obtain the optimum structure is left to his own ingenuity, at its best instigated by the salesmanship of an agricultural implement dealer and the catalogue of farm machines. In the consulting practice of the writer, about ten questions regarding fertilizer and humus are put to him as against one about soil structure. Yet the production of the right structure is as important for the success of farming as the chemistry and humus production of a given soil. In fact, without the proper physical structure no other soil treatment can become fully effective.

The ideal soil is crumbly. This means that its structure is porous, finer and coarser particles are equally mixed. Such a soil is penetrable by air and water. Water and air are important for the existence of microlife and the availability of soluble minerals. The crumbly soil, therefore, safeguards the maximum of microlife and the formation of natural humus. In a crumbly soil excess water filters downward, while in a tight, caked, crusted soil it runs off, thus causing erosion. In a crumbly top-soil the pores store up air for the breathing of the microlife and the roots. In fact 25 to 35% of the volume of such a soil consists of air.

The opposite of the ideal is a tight, caked soil with no or little air, reduced absorption and holding power of moisture. This type of soil will become acid, particularly with excessive moisture, and alkaline in the absence of moisture. It will tend to dry out faster under conditions of drought than the crumbly soil. Since we know that humus, in its best form, neutral, colloidal humus, absorbs and holds moisture, the effect of a crumbly soil is increased with increasing amounts of humus. When we speak of humus, we mean the digested (i.e. transformed by microlife) fraction of organic matter, not the crude organic matter itself consisting of roots, leaves, stubble, etc. Only a survey of the microlife and structure of a soil will inform us as to whether there is a change that humus formation can take place or not.

Ploughing, harrowing, disking, cultivation in general is done in order to turn and improve the surface structure, to crumble or powder the soil. Ploughing, especially, is a double-edged means, for it can not only loosen a soil, but it can also make a soil tight, creating a hardpan or a crust. Disking and harrowing also constitute a two-edged sword, for they cannot only loosen a soil, but they can also be carried too far and powder the soil too much. We must learn to differentiate between a crumbly soil and a powdery soil. A powdery soil is just as bad as a tight soil, it too will hold no water. When dry it is likely that the wind will blow it away as we have seen it happen in the dust bowl and after excessive listering. When wet it puddles and closes up at once, excess water will run off carrying with it the fine silt which colors the streams and rivers red, red-brown, brown or yellow. No water will penetrate into the depths and much of the rainfall needed to replenish the moisture in deeper layers will be lost. A powdery soil is dusty. A crumbly soil in general will not break down to dust, except during prolonged periods of drought. Soil bacteria and fungi will not live and survive as well in a powdery soil as in a crumbly one.

The ideal crumbly soil is one with a high humus content. It will maintain its moisture content to a high degree so that even after 4 to 6 weeks of drought crops will still grow. Corn and soya beans will germinate in a crumbly soil, even when it is moderately dry, while in a caked or powdery soil they will not.

A special case amongst wet soils is sticky, loamy soil. When wet, because of its powdery or caked surface, its particles stick together like glue. This is the most difficult of all soils to treat. Its response to fertilizer and manure is very slow. In fact, it requires many more years for its revitalization than any other kind of soil. Twice as much manure is needed here as in a crumbly soil to produce any effect. We have frequently encountered cases when people send soil samples to find out why, in spite of having abundantly fertilized and manured, their crops were unsatisfactory. The chemical analyses revealed that these soils were well provided with plant food, but that it was in their physical structure that they were deficient.

The chemical law of the minimum, which states that the mineral element which is least present (in the minimum) determines plant growth, can be broadened to include humus content, moisture absorption, moisture retention, and last but not least the structure of the soil. A structural 'deficiency' is as bad as a chemical deficiency.

The practical farmer who does not have a chemical, microbiological, biochemical and physical laboratory at hand can nevertheless use certain observations to be completely informed about the structure of his soil as any laboratory can tell him. These are the points to be watched:

1. **Water Behavior:** Observe during and after a rainfall whether the water is absorbed, runs off, or stands in puddles.

> *Perfect:* if after a heavy rain you can still walk over your soil without carrying the field away on your shoes.
> *Good:* the fields look moist but are not wet.
> *Fair:* the field is wet but after a few hours the glossy look has disappeared.
> *Poor:* puddles or runoff of water.
> *Danger symptoms:* the runoff water is clear but tastes bitter - minerals dissolved and washed out (horizontal, chemical erosion). The runoff water is colored - silt is washed away, your surface soil migrates to a location where it is of no use to you. These observations, by the way, are all important for correction irrigation.

Only the observations under *perfect* and *good* indicate safeguards for proper irrigation. If the poor or dangerous symptoms show up too much irrigation has been applied to an unsuitable soil.

2. **Moisture Retention:** Observe how long the soil remains moist without rain or irrigation, how long it will hold the water content. Here not only the surface appearance is important, but one should investigate how deep a rain penetrates, and how long the moisture lasts downward to 3", 6", 12", etc. How soon is the subsoil moisture consumed? Dig a hole with straight sides, a so-called soil profile and investigate.

Good: It is difficult to set the upper limit; a soil rich in humus may look dry and still have 25-35% moisture, while a powdery soil, looking just the same, may be bone dry. However, we would consider it good if, after 2 weeks of drought, a soil is still moist 1-2" below the surface, or if after 4 weeks of drought it is still moist 4" below the surface or if after 6 weeks of drought 2+ year-old alfalfa still grows.

Poor: If after the same periods of drought the soil is caked or powdery on the surface or dry underneath to the above mentioned depths.

Needless to say, a heavy plant cover and especially mulching is the best means of increasing the moisture retention. Mulch has the advantage that underneath it a crumbly structure remains or is formed, while although plant cover gives shade it does not always improve the structure of the soil.

3. **Moisture Penetration:** The structural, chemical and organic or mineralized state of soil has quite a bearing on the depth of moisture penetration. Layers of clay, especially of pure, blue clay are almost impenetrable. Moisture is dammed up above such layers and waterlogged soils result. Acid, raw humus with iron oxides and calcium carbonates, even small amounts of clay, will precipitate, wash downward and gradually form an almost impermeable layer. With acid humus and iron carbonates, one can find underneath old pastures, or forming in badly aerated and badly drained soils, a thin layer which is

impenetrable to plant roots, water and air. The writer was much impressed once when shown a dark, sandy, humus soil with very little clay content, which was completely water-logged to about 1 foot in depth; then there was a thin layer 1" thick underneath which one found completely dry, white sand.

Sometimes one can find such layers at various depths, where most of the lime, phosphates and other fertilizers applied over the years have been washed downward (vertical chemical erosion) and have accumulated. Every effort should be made to remedy this situation by means of subsoiling, even by surface dynamiting if necessary. These layers as well as any hardpan stop the growth of deep growing roots. On pastures and hayfields all deep rooting grasses and legumes, alfalfa being the most important, will gradually die and such fields will no longer be resistant to drought for there is no access to the deeper stores of water.

The aeration of the topsoil without forming a hardpan or acid humus, the avoidance of acid forming fertilizers, and good drainage are essential. It is a wide spread error to think that liming alone is able to cure such a situation. Lime with insufficient drainage, acid humus in a soil containing iron plus the lack of air speeds up the formation of such layers. The more lime is applied the faster the occlusion. The indiscriminating use of lime and fertilizer can do a lot of harm to the soil structure and render the expected benefits useless. The farmer should always keep in mind that excess damages are as bad as deficiencies. To obtain the proper balance, this is the wisdom and art of farming. Structural changes in a soil in the direction of hardening, crusting, puddling, etc. are danger symptoms of the first order and cannot be counteracted by fertilizing but only by increase of mature humus and proper tillage. Therefore, we shall now discuss a few points regarding tillage.

With the current high wages and the lack of sufficient farm help, as well as the sad fact that very few farmers know how to handle soils skillfully, more damage has been done by insufficient and improper tillage than by any other means. It should be the goal of tillage to obtain a crumbly aereated soil. The following outlines a few dos and don'ts.

Ploughing: *Never plough a soil when it is too wet.* When the clay sticks to the plough share and the bottom of the furrow gets that slick, glossy look, beware, the result will be lumps and hardpan. We needn't say: don't plow a dry soil for the surface will very likely be too hard to allow the plough to cut into it. Faulkner might not have needed to write about ploughless farming had the farmers observed this rule. The heavier a soil the more important it is to strike just the right amount of ploughing. The heavier a soil the more surface tillage it needs.

Never plough into the subsoil. Don't plough too deep. Only as much soil as will fall apart easily as it dries on the surface without lumping or caking should be ploughed or turned. The heavier and the more moist a soil, the less deep you should plough. If you are forced to plough an unsuitable soil and you can afford a top dressing of manure or compost, even a light one, before turning the soil, so much the better. But then, plough shallow. Never bury manure at the bottom of a deep furrow. If the soil conditions are not favorable, don't plough. The damage done by ploughing in the wrong way and the operations which are needed to correct a caked soil or hardpan cost more than waiting. Shallow ploughing goes hand in hand with organic methods. Consequently, subsoiling which cuts but does not turn becomes a must in order to obtain maximum results. Subsoiling breaks hardpans, brings air and moisture to the deeper strata, and increases the access of the plant roots to deep-seated moisture resources during droughts.

I believe that in well maintained organic soils ploughing can be reduced to a minimum. Instances where it is still needed are in the case of high stands of weeds and for turning grass sod on pastures and hayfields. With a heavy stand of tall weeds it will be better to mow them before ploughing, best (for those who can) to use the forage cutter, shred the weeds (or green manure crop) and blow or drop them back onto the soil, then plow it once. If you are turning under grass sods, the ideal will be to use a plough which makes a small furrow and turns the sod over completely. To wait for the right state of moisture when ploughing sods is important so that they will fall apart easily. Of course, this is easier said than done. The old-fashioned horse drawn sulky plough cutting narrow furrows did a better job in turning old, sticky, grass sods than most of the tractor ploughs. I am not in a position as yet to advise which plough bottom for the tractor plough would do the best job here. Before you plough under an old pasture or hayfield by all means subsoil first.

The most difficult problem is how to take care of a situation after a drought. Ploughing is out of the question. We have tried to break the hard surface with a corn cultivator or a spring tooth harrow. This works sometimes. If it works, it is worthwhile because the 'scratched' will absorb water faster than a caked surface and you will be able to work such fields at an earlier date than otherwise (for fall seeding for instance). Irrigated soils I would plough as little as possible. A disk plough might be handy, too.

Harrowing and Disking: Farmers in the old world have quite an array of different harrows, spike and spring type. Here we try to do most of this work with the disk. The disk harrow does an efficient and speedy job but sometimes it is difficult to produce a fine and even seed bed with it. Here the spike tooth harrow is the best means of obtaining a very fine surface structure. In order to level out a field, one of the best ways is to drag a log or rail over the rough surface, in spring, for example, over a fall ploughed field, before any other operation. This forces us to wait until the surface is dry enough and helps to prevent us from ruining the field by disking it too early (empiric education). In the old world this use of the log drag is a routine measure I believe most of the fine seed beds there are due to this practice.

My teacher in tillage was an old farm hand who had stayed on the same farm in Holland for 40 years. He knew his soil and was able to handle even the most difficult heavy, wet or dry soils, with success. He knew more about tillage practices than I could find in all the books here and abroad. One of his principles was: *the heavier and more moist a soil, the smaller and shallower the furrow.* Another: *When harrowing, work deeper with each operation.* That is, the first harrowing very shallow, the second a little deeper, the last deepest of all. He would harrow a field at least three times. To apply his principle to the disk, we may disk 2 to 3 times and then follow this with a long-toothed, spike-toothed harrow. Most of the modern tractor spike-tooth harrows are steel framed and have quite short teeth. They are all right for crumbly soils and surface work, but the old-fashioned wooden framed spike-tooth harrow with longer teeth does better, particularly in stony soils. A spring-tooth harrow will make a clean field and combat deeper rooting weeds. But be careful when using it on freshly ploughed-under sods and manure. It will bring lumps of manure and sods to the surface unless they have fallen apart completely. It will help, however, to dry out a soil.

The ultimate aims of tillage are to obtain a fine, crumbly seed bed, to keep fields free from weeds, and to prepare access to deeper layers of soil to the roots.

As far as root growth is concerned, it is important to have the soil equally loose or dense, not alternate layers of hardpan and loose soil. Roots will be handicapped or even stop growing when they strike a hardpan or lump. This causes considerable disturbance in the metabolism of plants. Living tissue will be undernourished for a certain length of time and thus becomes the easy prey of diseases and pests. Fruit trees and shrubs are particularly sensitive reactors to such differences in the root area. Leaf-roll and aphids develop much more readily and the protective cover of the epiderms (the waxy layer on growing grains, grasses and leaves) is weakened or disappears. Spraying cannot improve the inner soil structure and the farmer or gardener should not blame these things on 'poor seed' or 'ineffective sprays', but become conscious of his own errors in soil management which are often the true causes.

An important measure in the preparation of a seed bed is to give the soil a chance to settle between the different operations. Germinating seeds stretch their little roots out into the soil particles. If the soil is still in process of settling, these fine roots can break off and growth will be impaired, if not completely interrupted. Unless a soil is well settled and has been rolled or cultipacked, it is better to wait a few days, after spring ploughing, before disking. The exception would be when drought sets in so that surface lumps form, the dragging, disking or some other operation may be justifiable. I realize that the demand for speedy work in spring is a factor which limits waiting for the soil to settle. Another limiting factor is the local weather. In areas where the soil dries out rapidly one may not dare to wait too long during the spring planting season. Much knowledge of local weather and many observations essential to hitting upon just the right conditions. These are also valuable in deciding whether or not to roll. In theory, a fine, crumbly, or a particularly dusty soil should be rolled at once if there is any likelihood of drought, otherwise it will lose too much surface moisture through its capillarity. When a rolled or packed soil gets hit by a shower, however, the surface will remain closed and following drought will crust it or further rains will not penetrate. In many cases, then, the packed surface will require a very light harrowing with a short-toothed spike harrow or a weeder.

It is even beneficial to harrow (spike) a winter grain field just when it begins to grow in spring. This should not be undertaken, of course, until the soil is dry enough, and all danger of frost is passed; nor would one do it in the face of impending drought. The chief benefit of this practice is the breaking up of a crust which often forms after a wet, mild winter. A weeder is also suitable for this operation. A thick stand of grain, deeply rooted in a soil rich in humus, can bear such treatment. In Holland, with its mild winters, wet springs and firm, clay soils, it was my routine practice. Here, I have encountered difficulties, because of frequent, nasty droughts in March which delayed the operation beyond the favorable time, not to speak of the many stones in the fields. Yet, if this measure can be taken it will increase the grain yield considerably. It is the heavier soils, in particular, which need loosening up after rains.

By the way, rototillage produces a very fine, fluffy, billowing sort of seed bed which is apt to settle quite a bit and it needs time for this so that waiting is really essential here. A rototilled field which has suffered heavy rains immediately after cultivation is a sorry sight and usually must be tilled all over again. The careful weather observer will have the advantage here. Incidentally, such intervals between cultivation, disking, or harrowing, have another value. Weeds which were turned under may grow back to the surface, 'silent' seeds may be encouraged to germinate. These are all best combatted by the next cultivation just when they have come up again. Here too, impending drought is our worst enemy.

Now a word about field cultivation after harvest. If there is no grass and clover or other crop (green manuring, nurse, or cover crop) which had been sown into the grain, the field should be ploughed or disked at once after harvest. Every day of delay will harden the surface crust in summer, hinder penetration of the rains, and make it more difficult to establish an aftercrop. Unless the field is destined for a fall seeding of grain, it may be better to establish an aftercrop at once (soya beans, Sudan grass, millet, vetch, domestic rye grass). With favorable weather conditions, these crops can be pastured. Otherwise, they will be useful as green manuring and winter cover, particularly for land which has a tendency to erode. Fields with excessive weed growth or a lumpy structure may be dry cultivated (frequently disked) and manured for fall seeding. Late summer and fall are also the best time for subsoiling. After the harvest of grains the soil has lost its cover and hardens up very quickly. I remember instances when a field could have been tilled within the first three days after combining but after a week it was impossible with the conditions of drought which prevailed at that time.

Sometimes moisture conditions are still favorable in August. Then it pays to sow clover or alfalfa. A top dressing of earth compost with the last cultivation before seeding is excellent in such cases. It is especially important not to pasture all these new seedings at any time when the soil is too wet or even when it is merely moist. The trampling of the cattle leads to the formation of water puddles and to freezing out. Many a field has been ruined in this way before it could begin to bear fruit.

Old pastures and hayfields, especially those with a thick root felt, are best manured, limed if too acid, and then ploughed in late summer or early fall. The disintegration of the lumps of root and sod is favored by the microlife of the soil which reaches a peak in October, and the field will be in a more workable condition in the spring than when it is ploughed very late and freezes up right afterward.

Heavy clay soils are cold, hold winter moisture longer, and are usually the last ones ready for cultivation in spring. In planting oats, peas, or other early spring seedlings such soils may be the cause of much delay. It is preferable to have prepared them the preceding fall. Late summer is the time to subsoil, plough and till these fields. The same holds true for land which is to be reclaimed after drainage of a swampy or abandoned field. Such fields will be bumpy, uneven and very irregular. Buckwheat or Siberian kale may be put in as a cover to be ploughed under in late spring. After thorough disking, a planting of soy beans, field peas or beans, secures a very fine structure with which one can then go ahead with a normal crop rotation (manure and corn, for instance). Soy beans are also an excellent first crop for former old pastures. There are many species of 'silent' weeds which are awakened by the plough. Soy beans choke these out with less expenditure of effort in cultivation than would be the case with a cultivated corn crop. The corn crop which follows the soy beans will also be easier to handle. Short leys of two years of grain crops followed by two to four years of grass, clover, and alfalfa, are of course easier to handle than very old sods. These short term mixed grass and clover fields can be manured and planted to corn right away.

Light, sandy soils and those subject to erosion should always be covered with protective crops. Sandy soils are easier to handle in spring for they dry out and warm up more quickly. Cover and green manuring crops can be applied here right after harvest and be ploughed under in spring in plenty of time for early spring seedings. In general, where there is danger of erosion, or in mild southern climates, I would be inclined to maintain a cover crop over winter. Under moist more northerly climatic conditions it may be possible to leave the soil open in winter fallow. In humus soils with manure and/or compost a fall fallow favours the development of nitrogen fixing bacteria and it has been found that the nitrate content increases during the fallow period.

As a matter of principle, it should be understood that the humus effect of ploughed-under grass sods lasts for about two years and that these fields can be allowed to lie fallow for a longer period with less danger than can fields which have been bearing grain crops for several successive years. A field which has had three or more years of grains in succession should have a cover crop by all means. We repeat here an important fact which has been mentioned on other occasions. The enduring effect of manure, that is, the length of time which the microlife stimulating and humus forming effect of manure will last, depends mainly on two factors: the type of crop rotation and the climate (exposure to direct sun radiation). In hot, dry climates the organic matter is consumed much faster that in cold moist climates. Manure is also preserved longer in heavy soils than in light sandy soils.

The intervals between applications of manure must of necessity be shorter in hot climates than in cool moist climates. Therefore, manure may last only 4 to 6 months in the southern states unless cover crops are used with great intensity. In the middle latitude it will last about 2 years, in northern latitudes 2 to 4 years, in Canada about 3 to 5 years. Crop rotations alternating tilled crops and cover crops must be adjusted accordingly. In order to maintain the humus structure with a rotation of corn-wheat, corn-wheat, one would need a full manure dressing of 10 tons per acre every other year, which is practically impossible because it would require an incredible number of livestock. Only dairy farms might supply enough manure for such situations but these are not frequent in the typical corn-wheat belt. Inorganic fertilizer may boost a crop. The more organic matter there is in a soil, the better these fertilizers will perform and the less damage they will do. To believe that one can maintain a high humus level with inorganic fertilizers alone is a thought which still needs proof. A crop rotation management with short leys will stand up better under inorganic fertilizers than will the one-sided repetition of grain, grain, grain... However, one will require much smaller amounts of inorganic fertilizers if the soil has a good humus structure than otherwise.

There is no abstract principle which will suit every and all situations. One must observe the local conditions and then make decisions accordingly.

Bio-Dynamics, Volume VII, Number 3-4

XIII

SOIL PROFILES

A Diagnostic Means

Soil profiles can be very helpful in determining the overall structure of soils as well as in the diagnosis of the causes of various structural difficulties, the recognition of hard pans or other undesirable layers in the subsoil and deeper layers too. This paper does not deal with the rather complicated problems of soil structure, such as the aggregate state, analysis, classification of soils, etc., but is intended to enable the practical grower, farmer and gardener to get some basic information about his soils in a simple way, in a kind of field, spade analysis. The method is simple, the tools needed are few, but a great deal can be learned.

The Method:

Dig a hole, one foot square if you do not intend to dig deeper than one foot, larger if you wish to investigate deeper layers. It is important that two sides of the hole have sharply cut sidewalls (faces) and that the soil on these faces is not disturbed or compressed in digging and cleaning the hole. Once the hole is dug, peel a layer off the face without disturbing the structure. The best tool for this purpose is a flat spade, not a curved one; a spade used for ditch digging is ideal. This method will not work if the soil is so wet that water collects in the hole, nor so dry that it resembles rock. Neither of these conditions would permit the digging of a neat clean hole. Therefore, under the right moisture conditions, a neat hole can be dug.

The second tool to be used, after the hole is dug and a cleancut sidewall face is visible, is some kind of a scraper. A hand cultivator with three or five solid teeth (not spring or wire teeth) such as gardeners use, will do. In an emergency a dull, blunt knife or small spoon will work. Scrape or scratch carefully along the peeled face of the hole. Make horizontal strokes, beginning at the top; work gently and try to feel the structure of the soil, to determine whether it is crumbly, loose, tight, hard, etc.

Scratch parallel with the surface, slowly 'descending' so that you feel your way from layer to layer. Do not make vertical strokes. The idea is to 'feel' one layer after the other, to discover in this way the thickness of the various layers. Measure the thickness of each layer and plot the findings on a piece of paper.

Example:

Top inch	dry, dusty
1st to 3rd inch	loose, crumbly, friable, breaks easily
3rd to 5th inch	denser, still breaks easily, but not so crumbly
5th to 7th inch	very hard, caked
7th to 12th inch	slightly more crumbly, but not as good as top layer

Alongside the structure, note down the change in color and other physical properties, for instance, sandy, sticky, loamy, sandy-dry, or any other observations you may make. For annual crops it might suffice to go down one foot. For perennial crops, especially trees, the hole should be 2, 3 even 5 feet deep.

For those familiar with the terms A, B, and C horizon used in defining soil strata, it may be mentioned here that the layers found with this simplified technique are not always identical with the horizons, although the influence of cultivation (as was pointed to in the example given above) would belong to the A horizon. In order to avoid misunderstanding, we shall speak here only of layers or strata in terms of inches or centimeters.

Discussion of what can be found:

Structural changes: A healthy soil should have a crumbly structure, break easily when dry, and should not contain large lumps. The deeper this crumbly structure extends, the better the soil.

Heavy clay soil may tend to 'silt up' with fine, even colloidal material so that all the pores of the soil are closed. Such a soil gets waterlogged when wet and dries with a hard, caked surface. This crust or cake on the surface level is a severe handicap for successful agriculture. The best remedy is to incorporate more organic matter into the soil and increase soil life since microorganisms in soil have a very beneficial influence on soil aggregation (the formation of a crumbly structure). Mulching also has a beneficial effect, and such shallow rooting cover crops as will give enough shade to the soil so that a proper tilth can develop. A caked surface layer often results from working a clay soil while it is too wet or from using heavy equipment too much, i.e., packing the soil down. Seeds sown too deep in such a soil may have a difficult time breaking through the hard surface layer, if they do it at all.

Under the surface layer may be another, a little more packed but still friable. Then there may be a hardpan. This can be caused by plowing a clay soil when it is too wet so that it becomes compact under the plow bottom. Sometimes it is possible to find this glossy, polished plow sole long after it was first formed. If this compacting effect accumulates over many years, a real hardpan of considerable thickness forms. I like to call this the 'mechanical hardpan'. This cannot be penetrated by feeder, hair roots. The circulation of water and air is interrupted and the soil falls apart into 'independent' layers. The surface layer, down to the plowed hardpan is cut off from resources of moisture and minerals in deeper layers and becomes exhausted. Due to the lack of air in the hardpan and below it, no proper humus formation can take place and a shallow cultivated soil results. Deep rooted plants, like alfalfa, suffer particularly and freeze out easily because their roots scarcely penetrate the hardpan. Also they starve because the mineral and especially the trace mineral resources of the deeper layers are withheld from them. The shallow top layer will dry out readily due to the fact that it has no access to the deeper moist layers. Plants will suffer from even a short drought.

There are only a few sturdy weeds with tap roots which can break through such a hardpan. In fact, if a field has a great many such weeds, one can immediately suspect that it is a case of the survival of the fittest. Unless the farmer knows how to avoid this sort of hardpan, or knows how to cultivate the soil in order to remove it, it would be better to allow these weeds to do the job. Weedicides will certainly not solve the underlying cause.

The best mechanical means of avoiding the hardpan is not to work a clay soil when wet. Mechanical means of repairing the damage are to break the layer by deep cultivation with a spring or spike tooth cultivator or duck foot cultivator. The best tool of all, however, is the subsoiler. Plowing up and turning a hardpan on top, burying a crumbly layer underneath it, creates a soil filled with lumps. Many years ago the writer's attention was called to a field of sugar beets which had a very spotty look, large beets alternating with small ones. The latter were growing on lumps underneath, which had been created by too wet plowing, only the surface had been smoothed out by superficial cultivation.

According to the past history (i.e, cultivation) of a field, there may be one or several hardpan layers of varying thicknesses. An experienced observer of soil profiles can detect wrong cultivation several years back. Sometimes it can be seen that compost or manure had been buried by plowing too deep and wet, sealing it up between two layers of hardpan, indeed separating it from the top aerated layer, making the latter effective for shallow roots only. Such organic matter might lead to the formation of humic acids which in turn filter down with rain or irrigation into deeper layers. This either means complete loss or can give rise to such severe soil diseases as the formation of iron-alum-humic acid concretions which are very damaging to roots. A farmer will say that he applied a certain amount of manure or compost per acre on the surface of the field but the organic method did not work. It is well to remember that no organic fertilizer should be buried in or below a hardpan, unless the hardpan is broken by proper cultivation. The soil profile will show you where you stand and how to proceed.

It is obvious that the subsoil, which is of an entirely different structure and color, should not be worked into the top soil, or only to a small extent, namely as much as the top soil can take and 'digest'. Otherwise another type of lumpy pattern will be created, which is very damaging to the development of fine feeder hair roots, especially of plants which are not accustomed to growing into the subsoil.

The deeper layers below the actual cultivated level are of great interest as regards drainage, water supply or losses, standing water, and need study especially where shrubs and trees are to be planted.

Here are the two extremes. After a wet winter with not too much frost, a soil profile was dug in a pasture. The surface layer (18 inches) was squashy wet. The deeper subsoil, sandy loam two feet below, was bone dry. A hardpan between had sealed the subsoil and the whole winter's moisture had run off after the surface was saturated. Erosion under such circumstances is speedy and drought damages show up quickly. The other extreme is one where it is difficult to dig a hole for a soil profile because water oozes in from underneath and the sidewalls cave in. This can happen on badly drained land even though the surface may look dry. It is very important to find out where the water horizon is at the end of rains or winter as well as at the end of drought.

Knowledge of the water horizons is important especially if one is planting trees, the roots of which are sensitive to standing water (dying off or danger of fungus diseases). Only shallow rooters are possible for a situation with a high water level, where we might use those with low pyramid forms. The step to be taken when the deeper layers are permanently bone dry is to mulch the surface to train the feeder roots to grow near the surface, accessible to light rains or irrigation, that is to utilize the surface moisture.

Chemical Changes: Chemical changes are first seen in the color of the soil. Then samples of the different layers can be submitted for laboratory tests. The results of these tests are sometimes surprising and enlightening. There is, of course, the change from a black-brown humus soil to underlying yellow or red clay, white sand, yellow sand or blue or gray clay which is about as far down as agriculture goes. These changes are obvious. There are others, mostly of greyish shades which are not so easily observed. The color chart is manifold and sometimes deceiving. A black soil is not always a fertile humus. There is, for instance, the case of the blackish, wet clay, or clay-loam. Here is acidity, lack of air (oxygen) which, together with the reduction of organic matter, causes an increase of carbon and a loss of nitrogen, and a black 'staining' with humus acids or raw humus. These are not the right, black soils with 3, 4 or 5% organic matter, but frequently contain less that 1-1/2% organic matter. There is also the alkaline adobe soil which is also 'stained' black, but the color does not indicate a rich humus. Then there are the high organic matter soils tending to become peat. These abnormal cases can usually be easily recognized by their reaction (pH), rather acid or alkaline, and by their lack of a crumbly, friable structure. When wet, they are either spongy, sticky, or slimy; when dry, they cake, crust, crack in large crevices, or are dusty.

It is also of great help to determine the soil reaction (pH) in connection with the layers of the soil profile. The gardener, for instance, who had a lot of two-legged, split carrots and beets, found the subsoil to be very acid although the surface was all right.

Now for some of the surprises. There was a surface soil moderately low in nitrogen with its subsoil high in nitrogen. Here the farmer had applied a nitrogen fertilizer to a rather loose, light soil with little clay and humus in the top soil. There was no binding power and the rain (or irrigation) had washed the nitrate downward into the substrata and deposited it there out of reach of the roots of the annual plants on the field. It is true that such things may not happen too frequently with nitrogen, but they do happen to a rather large extent with phosphates, potash, and especially with calcium (lime). These, deriving naturally from the surface soils or fertilizers (or both), can be found way down in many 'chemical' layers of the soil. According to the permeability of the soil and whether the surface has enough 'binding' or 'holding' capacity, due to inorganic and organic colloids with a high absorption power or none, one can find much of the fertilizer of many seasons down below, inaccessible to all roots except the tap roots of weeds, trees or alfalfa. Especially do soils, one-sidedly cultivated with crop rotations of grains and no legume or grass cover rest periods between, show such phenomena. In dutch polders, the depth of the layer of calcium carbonate deposits below the surface indicated quantitatively the 'age' of the polder, namely, how long it had been under cultivation. This can be expressed in so many inches per decade. Phosphate fertilizer can be found far down. In general, one can speak of 'vertical chemical erosion'. It is well to investigate the deeper layers of the soil and bring about a correction so that the washing downward is counteracted. Some of these layers become very hard, cemented, and one can speak of a 'biological' hardpan. A good humus structure on top is one of the best remedies. In forest and prairie soils the washing downward of raw humus leads to podzol formation. All weathered products gradually collect in deeper layers so the surface is free of them. † This, however, means depletion of the agriculturally usable surface. The natural weathering process works in this way and the farmer helps it too.

A problem frequently encountered is the formation of iron-alumina-humic acid concretions with the subsequent formation of 'ortstein', or 'bog-iron', iron crusts. Yellow and reddish iron oxide containing pebbles, spots and discoloration are found on or near the borderline of top and subsoil, a little of it brought upward by cultivation, and extending downward into the substrata (B horizon).

In southern laterite soils this may even lead to the formation of real iron ore. In the soils of northern latitudes it is frequently found underneath badly drained old pastures (bottom land). These are serious disturbances in the chemical soil profile. Where raw acid humus prevails it supports this development, while a mild neutral humus does not produce 'ortstein' or 'orterde'. The proper humus management of the top soil therefore has an important bearing as a preventative and curative element.

Two anecdotes may conclude this article on soil profiles. One concerns the case of an orchardist in southern France who called the writer in for a consultation. A fruit orchard had been planted, but when the trees should have born fruit, they dropped their blossoms and a fungus infection began. From the pattern of the leaf discoloration and the dying of the tips of the branches, it was evident that root damage existed. Upon questioning, the orchardist explained that he had cultivated the soil very deeply before planting the trees and had fertilized heavily with sheep manure. The manure was actually three to four feet deep down, plowed in 'so that the tree roots would get something later on'. We dug a soil profile four feet deep and there was the sheep manure clearly visible, enbalmed after five to six years. Wherever a tree root had hit it, the disease symptoms showed up. This manure, by the way, was not B.D. composted.

The other story concerns the farmer whose land had been overused, overcultivated, and overfertilized for decades. It was run down and he decided to plant trees, to reforest it. Everything seemed to work out well for twenty-five years. Suddenly the entire forest became fungus infected and after thirty years died off. A soil profile was dug and showed that the trouble started when the roots hit a layer of chemical erosion, which had been created when the land was under cultivated crops many years ago and had remained as a stratum which was poison to the tree roots.

† An excellent book on the problems of soil horizons and podzols is *SOILS: THEIR GENESIS AND CLASSIFICATION*, by C. F. Marbut, published by the Soil Science Society of America, 1951.